Improving Energy Efficiency in Apartment Buildings

Improving Energy Efficiency in Apartment Buildings

John DeCicco, Rick Diamond,
Sandra L. Nolden, Janice DeBarros,
and Tom Wilson

American Council for an Energy-Efficient Economy
Washington, D.C. and Berkeley, California
1995

Improving Energy Efficiency in Apartment Buildings

Published by the American Council for an Energy-Efficient Economy, 1001 Connecticut Avenue, N.W., Suite 801, Washington, D.C. 20036 and 2140 Shattuck Avenue, Suite 202, Berkeley, California 94704

Cover art copyright © 1995 M.C. Escher Heirs/Cordon Art—Baarn—Holland. All rights reserved.

Cover design by Chuck Myers
Printed in the United States of America

Library of Congress Cataloging-in-Publication Data

Improving energy efficiency in apartment buildings/ by John DeCicco, Rick Diamond, Sandra L. Nolden, Janice DeBarros, Tom Wilson
     p.301   23 cm.
     Includes bibliographical references and index.
     ISBN 0-918249-23-6
     1. Apartment houses — Energy conservation. I. DeCicco, John M., 1952 -.
TJ163.5.D86I46 1995   95-47067
696—dc20   CIP

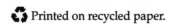 Printed on recycled paper.

Acknowledgments

This book rests largely on the dedication and hard work of a diverse community of individuals who have strived to save energy in apartment buildings for more than two decades. We cite many for their published work; others assisted by supplying information, answering our questions, and reviewing drafts of the manuscript. But the specific citations leave out a multitude who helped build the body of knowledge and evolved wisdom that is compiled here. Thus, we owe a great debt to everyone who has toiled in the challenging area of energy conservation in multifamily housing and dedicate this book to their past accomplishments, whereby energy and dollars were saved and the quality of apartment life improved in many places throughout the country. Much more remains to be done to bring the overall stock of apartment housing in the United States up to better standards of energy efficiency. We hope that this volume will serve those who continue to work for progress in this field.

We are enormously grateful to a number of people who quite tangibly helped bring this book to fruition. Steve Morgan, formerly of Citizens Conservation Corporation (CCC) and now with EUA/Citizens Conservation Services, Inc. (CCS), was instrumental in the project from start to finish, helping us develop recommendations for Chapter 6 and providing wisdom throughout the process. Other CCC/CCS staff, particularly Lillian Kamalay, now at CCS, and John Snell of CCC, also contributed key information and insights; Andrea Geiger and Theodore Lubke assisted in drafting material on apartment building conservation programs while working as interns at CCC. Loretta Smith, formerly with ACEEE, drafted the Chapter 3 material on improving the efficiency of water heating, appliances, and lighting.

We are very thankful to individuals who provided information that enhanced our research and reviewed drafts of the material: Roger Colton, Howard Geller, Robert Groberg, Blair Hamilton, Martha

Hewett, Norine Karins, John Katrakis, Larry Kinney, Paul Knight, Paul Komor, Jeff Lines, Mary Sue Lobenstein, Mike MacDonald, Mike Mc-Namara, John Morrill, Mike Myers, Steve Nadel, Andy Padian, Larry Palmiter, Scott Pigg, John Proctor, Miriam Pye, Len Rodberg, Beth Sachs, Debra Okumo Tachibana, Ed Vine, Mike Weedall, and Jack Woolams. We also thank Renee Nida and Mary Anne Stewart for their expert assistance in editing and copyediting and Glee Murray for her supportive management of production. Major support for the research and publication of this book was provided by The Energy Foundation and the U.S. Department of Energy, Office of Building Technologies.

Foreword

Owners and managers of apartment buildings have always had more pressing concerns than worrying about the energy efficiency of their properties. Filling vacancies, collecting rents, replacing broken windows, fixing plumbing problems—these and many other basics of maintenance and upkeep often take a higher priority than energy-saving improvements such as retrofitting ancient heating and hot-water systems, replacing inefficient lighting, and thermally upgrading the building's envelope. Yet these latter issues can have a big impact on a property's bottom line. Decades of underinvestment in energy efficiency have left behind a wealth of energy efficiency opportunities, cost-effective measures that can generate dollar savings of 30–40% and more on utility bills. Standing between the efficiency provider and the property owner is a minefield of obstacles: split incentives between the owner who pays for the equipment and the renter who pays for its energy usage; the requirement that the maintenance person, site manager, and financial officer must agree to recommend to the owner the efficiency improvement; the poor understanding of efficiency technologies and their performance; poor cash flow and credit histories; expectations of a quick payback for any money an owner needs to spend; and myriad competing priorities.

Overcoming these obstacles requires an understanding that there are not one but many markets that mediate energy use and efficiency investments in apartment buildings. There are public, publicly assisted, nonprofit, and private owners; there are individual, partnership, corporate, and institutional ownerships; there are low-rise, mid-rise and high-rise buildings; there are electric, gas, and oil-heated buildings; and there are differences in type of occupancy as well as regional differences in many factors. Owners in each category have different concerns, time horizons, and priorities.

Yet building owners do replace furnaces, hot-water systems, refrigerators, and other energy-intensive appliances all the time—when they break down, pose safety problems, or cost too much to maintain. Most owners even budget annually for capital improvements to pay for these items. In selecting new and replacement equipment, building owners usually put lowest cost, ease of ordering replacement parts, and durability at the top of their list. Unfortunately, energy efficiency—or lowest lifecycle cost—is often not even on the list. But it is at the point of this equipment replacement decision—too often made in an emergency situation—that the greatest opportunity presents itself to whisper into the ear of the owner, "The efficient option is the best option." Successfully getting the attention of a building owner— or of a property manager, in the case of large management companies—means making the case that lower operating costs will make for a more competitive property. And this case must be made in market circumstances that are often trying, with pressures to meet housing demands of households with stagnant or falling real incomes and now yet further pressures of decreasing public resources.

The proposed "reinvention" of public and assisted housing throws one-third of the existing stock of buildings with five or more units into turbulent waters. Housing vouchers would replace operating subsidies to public housing authorities (PHAs), leaving PHAs to compete with private housing for low-income tenants. The collapsing of scores of housing programs into two or three block grants would leave fewer vouchers to support assisted housing properties and housing authorities. These threatened cutbacks in turn stimulate owner interest in lowering operating costs to attract tenants in a more competitive marketplace.

Whatever the type of ownership, lowering utility costs is one of the best opportunities for lowering operating costs in apartment buildings. But the realization of these opportunities will not be solely market-driven. Electric, gas, and oil prices have been stable or slightly declining in most areas of the country, so that only rapidly increasing water rates have attracted concern in recent years. Improvements in low-volume flush toilets and more effective valve retrofits for existing toilets—in conjunction with leak repairs—offer savings of 30–40% and higher. Peak-period pricing and demand charges have created the need for greater control over the operations of heating, hot water, pumps, and lighting systems, which in turn has led to increasingly more effective and less expensive energy management systems appropriate for mid- and high-rise buildings. Utility demand-side-management (DSM) programs have introduced emerging technologies in lighting for common areas in apartment buildings that have cut usage by 30–50%.

Although spending for conventional DSM programs is in decline, overall opportunities have not declined since many utilities now recognize the importance of DSM programs and are refocusing their efforts on the more cost-effective approaches. Utilities have come to view DSM not only as part of a least-cost energy mix but also as an important customer service that can be valuable for attracting and retaining customers in a competitive environment. New DSM approaches include demand-side bidding for "blocks" of conservation and peak load reduction, such as a recent arrangement to cut 1.5 megawatts of demand through installation of more efficient equipment in a set of California apartment buildings. Although the emergence of retail competition is stimulating interest in bulk purchasing of energy supplies (both electric and gas), the principles of integrated resource planning remain important for ensuring that energy service needs are met equitably and at least societal cost.

Because they represent large customers, apartment buildings, and particularly public housing complexes, are well situated to take advantage of opportunities for investments in efficiency upgrades as a way to competitively contribute conservation resources as part of a diversified, least-cost energy mix. Cooperative purchasing networks—extended to address demand-side as well as supply-side resources—provide a structural framework for realizing these new opportunities. For example, the New York Power Authority recently formed a partnership with the New York City Housing Authority to stimulate the design and manufacture of a super-efficient refrigerator for the apartment market, providing an exciting precedent for the development of very efficient appliances for the multifamily sector. Other housing authorities will be invited to piggyback their refrigerator purchases on top of New York's later in 1996.

The New York Power Authority/New York City Housing Authority partnership includes a second feature that may have even more long-term significance: the utility is financing the purchase of the refrigerators, enabling the Housing Authority to pay back the loan from the annual savings generated. This concept, called energy performance contracting, is a little-known and less practiced financing instrument for the multifamily sector. Because it provides one-stop shopping and off-balance-sheet financing for building owners, performance contracting could provide a powerful incentive for the owner conditioned today to buy equipment only when existing systems are failing. Because performance contracting can incorporate long-term maintenance and resident cooperation in the operating of energy systems, it addresses two vital factors in the long-term persistence of savings.

Energy performance contracting has made recent inroads among public housing authorities, assisted by a recent regulation facilitating its use and boosted by the specter of competition with the private sector. Assisted-housing owners, stymied by Department of Housing and Urban Development (HUD) regulations governing debt financing and general contracting that are inappropriately applied to energy retrofits, can look forward to relief afforded by Fannie Mae's entry as an investor in the energy efficiency financing arena. With the assistance of a private energy efficiency financing firm, Fannie Mae is working to remove the contracting and financing barriers. Fannie Mae's willingness to offer performance contract loans and other energy financing products to the residential sector—with an emphasis on apartment properties—brings visibility, credibility, and confidence to the building owner and energy services company.

The slow pace of efficiency investments in the multifamily sector reflects the extent of the market and institutional barriers referenced earlier and underscores the need for continuing governmental involvement in the field. At the federal level, there is growing interest in this building sector among the Department of Energy (DOE), HUD, and the Environmental Protection Agency (EPA). DOE's financial support for this book, the first comprehensive examination of the field in almost 15 years, is timely. The authors of this book—experts drawn from a range of engineering, academic, and financial disciplines—bring the economic, technical, and policy issues to life.

At a juncture when the Congress is focused on shrinking the federal government's role in housing, energy, and other sectors, simplifying the regulations that govern assisted and public housing transactions, thus allowing the creative aspects of market to operate, is welcomed. Yet we must recognize a continuing need to level the playing field for energy efficiency choices in this particularly challenging building sector. Tax policy, R&D, performance mandates, technical assistance, evaluation, and other government roles are important. This book provides technical, programmatic, and policy information that can inform and stimulate discussions that are long overdue. It should also inspire action from each of the stakeholders in the field. That is the challenge for all of us working to bring energy efficiency to buildings that desperately need the attention.

—*STEVE MORGAN*

Contents

Introduction

Apartment buildings with five or more units per building account for 9% of residential energy end use in the United States. Most apartment dwellers are renters. Some live in multifamily housing* for most of their lives; others use apartments as a stepping stone on the way toward a single-family house. In any case, apartment buildings are a crucial part of the affordable housing stock and will remain essential for sheltering millions of Americans. The variety and complexity of apartment buildings, the generally low-income setting—including much public housing—and the fact that most units are rented rather than owned all contribute to the difficulties of addressing any issue in this sector, be it energy use, state of repair, security, or preservation. These issues are, in fact, intertwined.

Apartment buildings have long been identified as a particularly challenging area for energy conservation. There are technical unknowns about what the best retrofits are and how to implement them. Financial motives for conservation are often absent because of the split-incentives problem: tenants have no interest in investing in efficiency improvements because they do not own the building and may have short periods of occupancy; landlords may not invest in efficiency improvements because they can generally pass energy costs on to tenants and retrofits often appear risky or unprofitable. Many building owners and managers lack information about the potential

---

* Except where otherwise indicated, the terms *apartment building* and *multifamily housing*, which we use interchangeably in this book, refer to residential buildings of five or more units.

1

savings of energy efficiency improvements, their implementation, and the available financial assistance and incentives for such improvements. Landlord-tenant mistrust is a perennial problem. Other institutional barriers include poor training and lack of technical expertise among building staff and trades relating to energy-using systems. Tax structures may allow deduction of energy costs but require depreciation of energy-saving improvements. Obtaining assistance in paying utility bills is relatively straightforward, but procedures for securing retrofit financing are more complex.

Addressing energy consumption in apartment buildings combines the more challenging aspects of energy conservation in single-family residences and energy conservation in commercial buildings. Apartment buildings are physically more complex than single-family houses. The institutional milieu in which decisions are made, the building is managed, and financing is obtained is also complex. Most apartment buildings are commercial properties, and publicly owned buildings share some characteristics of their commercial counterparts. Like commercial buildings, apartment buildings involve a multiplicity of actors in managing the building structure, energy use, and energy-using equipment. On the other hand, apartment dwellers are still individual households who make their own decisions about how they live in their apartments while generally facing imperfect information, limited access to capital, and tight personal budgets. Since many apartment dwellers rent, they have few of the financial and tax incentive mechanisms that homeowners and commercial tenants have at their disposal to help manage building-related expenses, such as energy bills and the purchase and maintenance cost of energy-using equipment.

In the early 1980s, a number of studies identified the opportunities for, and barriers to, energy conservation in multifamily housing (OTA 1982; Bleviss and Gravitz 1984). Since then, much work has been done on both the technical and the programmatic aspects of the issue. Although more remains to be learned, the efforts of local, state, utility, federal, and nonprofit programs have provided valuable experience regarding effective approaches to improving energy efficiency in apartment buildings. The Department of Energy (DOE), the Department of Housing and Urban Development (HUD), and a number of state and city organizations have developed manuals and guidelines for audit and retrofit of multifamily housing. Numerous papers and reports on many aspects of the issue have been published and presented in the American Council for an Energy-Efficient Economy (ACEEE) Summer Studies programs and other forums. However, to date there has been no compilation and distillation of the accumulated experience and lessons learned from the past decade's efforts in this field.

Looking back over nearly two decades of effort in energy conservation reveals an evolution in the motivations and methods for addressing energy use in apartment buildings. Initial efforts had a sense of crisis. Apartment dwellers were perhaps the most poorly equipped to deal with the rapid rise in energy costs precipitated by the 1973 oil embargo and the 1979 Iranian revolution. The predicament was compounded by a lack of knowledge about how buildings worked and how best to improve their energy efficiency. Although not unique to the sector, this lack of knowledge was perhaps most acute for apartment buildings and, despite some progress, remains a problem to this day. Limited by a lack of technical know-how, early efforts were marked by prescriptive programs. Conservation efforts often focused on the building shell, emphasizing storm windows or double glazing, weatherstripping, caulking, and maybe tuning up the boiler. The field abounded with horror stories of energy waste: costly fuels inefficiently burned in decrepit boilers delivering uncontrolled heat that was then spewed from open windows (sometimes expensive, brand-new open windows) in a mostly overheated building. Practitioners in the nascent field of building energy conservation had to climb a steep learning curve, especially with apartment buildings, which are more complicated and which received proportionately fewer energy conservation resources compared with single-family buildings.

A three-story apartment building on Bosworth Street in Chicago, where "walkup" units are a large part of the apartment stock.

3

By the late 1980s, the stabilization of energy prices had taken away the sense of crisis, but our appreciation of the depth and breadth of energy use issues in multifamily housing had grown. We now see that apartment buildings are complex systems, involving a set of interrelated physical components that interact with the residents, staff, and management of a building. The institutional context, including such issues as private versus public ownership or assistance, housing affordability, and preservation, is also a major determinant of what can be done and how the work can be financed. At the same time, however, our collective confidence has grown. Less effective approaches to conservation, pursued with good intentions but lack of knowledge, can now be avoided. A number of energy conservation success stories—some large, some small—have all indicated that we can do much to cost-effectively upgrade apartment building efficiency. We now have examples that show how energy efficiency improvements can often enhance comfort, quality of life, and affordability, as well as save energy. For some types of apartment buildings in a number of regions, the energy conservation field can rightly say, "We know how to do it." Although many buildings still need attention, the constraints are largely ones of time and resources. Nevertheless, many challenges remain. Good information and confidence are not available for every region or every type of building. For example, successful approaches to energy conservation for more recently constructed buildings are much less documented, especially in the Sunbelt, where cooling (and sometimes dehumidification) are as important as heating.

In this book, we examine approaches to effectively reducing energy consumption in apartment buildings. We hope to serve a variety of professional audiences that have an interest in apartment building energy issues. For local housing officials, utility conservation program managers, and housing advocacy organizations, the book offers guidance in how to make successful investments in upgrading apartment building energy efficiency. For owners and managers of private and public multifamily housing, the material presented here should build confidence in the workability and benefits of investing in energy conservation, as well as provide examples of effective financing, performance contracting, and management approaches. For practitioners in the heating and air conditioning professions, energy service contracting, and construction trades, we present information needed to better understand the special needs of, and the opportunities for serving, the multifamily housing sector. Researchers will find an up-to-date summary of the field and indications of areas needing future work. Finally, we address policy makers by discussing program successes, deficiencies, and opportunities for improvement.

4

The City Center Plaza in Oakland, California, provides over 300 units of housing for a mostly minority population. The building mixes condominiums and rental units.

The Sector in Brief

One common image of multifamily housing is of older, high-rise apartment complexes. In fact, the U.S. multifamily stock is mostly less than 25 years old, with many low-rise, walkup, garden-apartment, or motel-style buildings. Apartment buildings contain 14 units on average. Buildings of 5 or more units comprise 21% of housing in the Northeast, 19% in the West, and 14% in the South and Midwest. Larger buildings tend to be in the Northeast, where 42% of buildings having 50 or more units are located. The South has the largest share (34%) of smaller apartment buildings (5–9 units). Multifamily buildings are located 53% in cities, 42% in suburbs, and only 5% in rural areas. Although the geographic statistics are probably expected, the age statistics might be surprising. More than half of multifamily units are in structures built since 1970; less than 20% were built prior to 1950. The past two decades' building booms in the South and West greatly swelled the numbers of multifamily buildings in these regions, which became overbuilt in some areas by the late 1980s.

Perhaps the single most defining characteristic of apartment households is that they are low-income renters rather than owners. Only 10% of apartment dwellers are building owners or condominium owners. The remaining 90% are renters, versus 30% of all U.S. households. Apartment households have lower-than-average incomes—over 5 million are eligible for means-tested federal assistance. Compared with households in single-family dwellings, apartment households spend higher fractions of their incomes for both housing (rent) and energy bills. Low-income apartment residents face very tight household budgets, often having to choose among paying rent, food, medical, and energy bills. Although the ratio of single family homes to apartment buildings has been stable nationwide, low-rent multifamily housing stock is decreasing. In particular, subsidized multifamily housing is being lost through disinvestment, abandonment, or conversion to higher-income units. Rehabilitation efforts that include energy conservation improvements are an important way to help preserve affordable housing.

Barriers to Retrofit Activity in the Multifamily Sector

The historically low level of retrofit activity in the multifamily sector has been due to a number of complex and interrelated barriers (Bleviss 1980). Although there are no simple solutions, understanding these obstacles and identifying the program design and research needed to overcome them are essential in furthering energy conservation for this sector. These barriers fall into several general categories: technical, informational, legal and regulatory, the "split-incentives" problem, and other economic and institutional barriers (see Table 1-1). The following discussion of the barriers to energy efficiency in apartment buildings is based on Diamond et al. (1985, 21–23).

Technical Barriers

Technical problems in retrofitting apartment buildings tend to be more complicated and harder to solve than technical problems in single-family residences because the multifamily stock is sometimes in poorer condition and the factors influencing energy use are more complex than in the single-family stock. Such technical uncertainties inhibit building owners from investing in retrofits without guaranteed savings. Retrofit installers cannot afford to undertake the necessary research to demonstrate the performance of their products. This technical barrier is exacerbated because research and retrofit testing can be complex and costly for apartment buildings.

Technical expertise on apartment retrofit performance is less available because technical efforts to date have concentrated on the larger single-family and commercial building markets. Retrofits in single-family buildings tend to be shell dominated, concentrating on insulation and storm windows, whereas retrofits in commercial

Table 1-1

Barriers to Energy Efficiency in Multifamily Housing

Technical Barriers
- Complexities and uncertainties that complicate audit and retrofit planning
- Lack of confidence in retrofit performance and savings
- Installer inexperience with certain building and equipment types
- Health and safety issues

Informational Barriers
- Lack of data about building stock
- Uncertainties about conservation measures already installed
- Limited and sometimes conflicting information about retrofit cost, performance, and behavioral factors
- Weak system for disseminating the information that is available
- Poor information on financing options
- Few knowledgeable people

Legal and Regulatory Barriers
- Existing building code violations
- Rent control without pass-through of beneficial improvements
- Utility rate structures
- Poor enforcement and evaluation of building codes

Economic and Institutional Barriers
- Split-incentives problems
- Tenant/landlord relationships
- Property managers' incentives
- Access to capital and lender reluctance
- Neighborhood deterioration and changing land use patterns
- Poor market valuation of "hidden" efficiency improvements
- Unequal tax treatment of energy efficiency investments versus energy consumption expenses
- Metering practices and questions of cost allocation among adjoining units
- Security and privacy concerns
- High tenant turnover

buildings are usually system dominated, emphasizing heating, ventilation, and air conditioning (HVAC) and lighting retrofits. Apartment buildings fall somewhere between the two, and because the demand for services from this sector has been less, contractors and energy auditors have had less experience with these buildings. Health and safety issues, ranging from fire codes to security concerns, also serve to complicate the technical aspects of apartment building retrofit.

Informational Barriers

Informational barriers to energy conservation in apartment buildings exist at several levels. Few detailed data concerning the regional multifamily stock are available to agencies and policy makers attempting to plan programs to further energy conservation. Limited and conflicting information about the performance, costs, and behavioral factors regarding retrofits prevents building owners from making informed decisions. The lack of reliable, credible information about financing programs—particularly alternative financing methods—and about the reliability of companies and practitioners is another obstacle. Reliable information—regarding particular building types, not just apartment buildings in general—is required to motivate retrofit investments. Compounding the information limitations is the general inadequacy of effective information transfer. Although owners and managers of large properties may read trade journals on building management, the large percentage of small-building owners are not adequately informed of what can be done to improve their buildings and how to go about it. Not surprisingly, there is a limited nationwide pool of experienced individuals with expertise on improving apartment building energy efficiency.

Legal and Regulatory Barriers

A hard-to-measure and little-discussed barrier to apartment building retrofit is the presence of building code violations and other illegalities. Owners of buildings with building code deficiencies, unassessed and possibly unauthorized improvements, or altered meters have reason to fear that energy auditing and retrofit activity will lead to discovery and penalty. Unfortunately, it is the stock with the greatest need for retrofit that is most likely to have building code deficiencies.

Other legal and regulatory barriers include rent control laws that do not have provisions for rent increases reflecting the value to tenants of energy efficiency improvements, legal restrictions on metering individual apartments, and regulations that allow utilities declining-block-rate structures that artificially make such metering uneconomic. Poor enforcement and lack of evaluation are obstacles for regulatory measures intended to improve the stock. For example, little is known about the effectiveness of requirements to bring buildings up to code at time of ownership transfer.

Rick Diamond

A large high-rise complex in Philadelphia, where apartment buildings provide housing for a wide range of income groups.

Split-Incentives Problems

Historically, a major barrier to retrofits and energy conservation in rental properties such as apartment buildings has been the split in economic interest between landlords and tenants. On one hand, if the tenants pay for energy, the landlord has little incentive to make physical improvements to save energy. On the other hand, when the landlord pays the energy bills, tenants have little incentive to control usage. Split-incentives conflicts between landlords and tenants occur in the general context of landlord/tenant relationships, which often entail mutual distrust. This fundamental conflict makes cooperation with respect to energy conservation particularly difficult.

Split incentives are often complicated by the independent interests of building managers, who do not pay energy costs but who wish to lower the effort involved in building management. Unlike owners or tenants, property managers tend to be isolated from paying energy costs. They define their jobs as working to minimize tenant and landlord complaints. A tenant complaint about lack of heat is more likely to be addressed by turning up the furnace than by insulating the apartment. Property managers may be reluctant to approach landlords with proposals for energy retrofits lest they appear as having failed in their maintenance duties or as adding to owners' expenses.

Other Economic and Institutional Barriers

Lack of access to capital for improvements is also a major barrier to apartment building retrofits. A substantial number of apartment building owners are financially strapped. Another complication involves lenders, who receive little benefit from successful retrofit projects but may incur financial risks associated with unsuccessful ones. Lenders are therefore often reluctant to absorb the apparent risks, especially long-term risks, of financing efficiency upgrades in apartment buildings.

A large number of apartment buildings are approaching the end of their economic life. To be financially attractive, improvements to buildings that may amount to a major rehabilitation must pay off more rapidly than improvements to otherwise similar buildings. Thus, building and neighborhood deterioration and changing land use patterns all act as barriers to retrofitting.

To the extent that the rental housing resale market does not reflect the value to the owner of energy conservation improvements, this market failure is also a barrier to energy conservation since it limits the liquidity of the owner's retrofit investments and discourages such investment.

Under the U.S. tax code, landlords may deduct the annual energy costs of their properties from their income as business expenses. They receive no parallel benefit for investments in energy conservation. In addition, concern about property tax increases based upon reassessments triggered by energy efficiency improvements may also deter such improvements.

A steam-heated walkup apartment building typical of the Minneapolis and St. Paul area.

Metering is another important institutional barrier. When buildings are converted from master meters, where the owner pays bills, to individual meters, where the tenant pays, the resulting reductions in energy costs are not always proportional to the reductions in energy usage because of rate structures that may favor single large users. Under these circumstances, owners may retain the master meter, install check meters for each unit, and bill the tenant for their usage. This practice is illegal in certain states that prohibit individuals from "reselling" fuel and electricity, the rationale being that unscrupulous owners could charge more for the energy than they are paying.

In metering conversions, the physical characteristics of apartment buildings pose additional complications. For example, one unit may be maintaining low thermostat settings to "borrow" heat from adjacent units. An interesting question about metering conversions is the impact they have on future retrofits. Potentially cost-effective measures, such as solar hot water and district heating schemes, are all more amenable to central space conditioning and domestic hot-water systems. Perhaps the most significant problem concerning metering conversions is that they are widely regarded as a panacea for apartment buildings in spite of their limitations.

Concerns about security and privacy can make both tenants and landlords reluctant to have buildings open to strangers associated with retrofit activities. Fear of vandalism—whether aimless or a purposeful part of landlord/tenant conflict—can act as a barrier to efficiency improvements potentially subject to such abuse.

Finally, the high tenant turnover that occurs in some apartment buildings serves to inhibit installation of energy-saving equipment in individual units. More efficient lighting and appliances, as well as devices such as setback thermostats, require understanding and care by residents. Landlords can be reluctant to invest in unit-level improvements because of concern about misuse or abuse by new tenants.

An Overview of the Book

Further details on the composition of the multifamily sector are covered in Chapter 2, which provides a broad set of statistics characterizing five-or-more-unit apartment buildings. Chapter 2 also addresses the institutional make-up of the sector in terms of who owns and who is responsible for operating, maintaining, and upgrading the buildings.

Chapter 3 looks at the technical and behavioral aspects of apartment building retrofits. Although lack of reliable technical information—what to do to improve energy efficiency—has been an obstacle to efficiency improvements in the multifamily sector, we can now trace the substantial progress that has occurred since the early 1980s. An emerging theme is the importance of treating a building as a system, reflecting interactions among physical retrofits as well as the need to educate and involve building management, tenants, and staff. This holistic approach to retrofit provides substantial energy savings but requires substantial investments and follow-up. Case studies from the Northeast, Midwest, and West Coast regions highlight different technical solutions to apartment building energy conservation problems.

Chapter 4 covers the programmatic aspects of implementing energy conservation in the multifamily sector. We review the history of multifamily conservation programs, along with lessons learned from past programmatic experience. Program marketing, information provision, technical assistance, performance standards and ratings (mandatory or voluntary), and financial incentives must all work together to significantly improve energy efficiency in multifamily housing. Monitoring, program evaluation, and follow-up are needed to ensure that savings are achieved. We highlight a number of successful programs from around the country that provide models of how to effectively implement energy efficiency improvements.

Chapter 5 deals with the financing of energy efficiency improvements in apartment buildings. We introduce the primary financing mechanisms, including grants, loans, bonds, tax-exempt instruments, tax credits, leases or lease-purchases of efficient equipment, and energy performance contracting. A persistent theme is the poor financial condition that afflicts much of the nation's affordable multifamily housing stock after more than a decade of malign neglect exacerbated by the meaner aspects of federal policy in the 1980s and 1990s. New opportunities for financing retrofit and rehabilitation in the multifamily sector are discussed. In conjunction with a stronger commitment and public and private (such as utility) resources to finance multifamily conservation, special efforts are needed to educate building owners, housing authority officials, and prospective lenders about the opportunities for successful financing of energy efficiency improvements in apartment buildings.

The book closes with a chapter on recommendations for accelerating the rate of energy efficiency improvements in the multifamily sector. Such prescriptions must consider the diversity of the multifamily sector in terms of physical building type, fuel use, geography, type of ownership, and the variety of institutions involved. We list actions that can be taken by major institutional actors, federal agencies, state and local agencies, and utilities. Chapter 6 also identifies the research needed to continue advancing our knowledge of how to improve apartment building energy efficiency.

Three appendixes provide supplemental information. Appendix A defines acronyms used throughout the book and common in the energy conservation field. Appendix B lists institutional resources, including government offices, businesses, and other organizations having expertise in apartment building energy efficiency improvements. Finally, Appendix C discusses specialized tools, including computer software, for auditing apartment buildings and planning retrofits.

Chapter Two

An Overview of Apartment Housing in the United States

Over 30 million Americans live in apartments—what we term multifamily housing. A disproportionate number of them are poor, renters, minority, single parents, and children. Because so many apartment households are located in the nation's urban centers, housing problems are often intertwined with social and economic decay. And although the dream of owning a single family home has motivated generations of Americans, apartment buildings will continue to provide housing for millions of Americans.

The multifamily sector covers a wide range of building types, from duplexes and low-rise garden apartments to high-rise structures occupying entire city blocks. Because of this variety of building types, no single criterion best categorizes multifamily housing. Previous attempts to identify this sector have used such features as number of stories, heating system type, rented versus owned, and number of units. However, as noted in Chapter 1, we here define the multifamily sector to be those buildings with five or more units, following the usage in a principal data source for this chapter, the Residential Energy Consumption Survey (RECS), compiled by the U.S. Department of Energy (RECS 1990a, 1990b, 1990c). Thus, this book targets medium-sized and larger apartment buildings. Again, as noted in Chapter 1, except where otherwise indicated, the terms *apartment building* and *multifamily housing* are used interchangeably in this book to refer specifically to buildings of five or more units.

The multifamily sector differs from the rest of the nation's housing stock in certain key respects: geographic distribution of units, age of

15

buildings, ownership patterns, utility metering type, and heating system equipment, among others. A detailed discussion of these characteristics, both physical and demographic, helps to identify the significant retrofit potential of apartment buildings and also highlights some of the key barriers that limit energy efficiency investments in this sector.

Characteristics of the Multifamily Sector

Size of the Multifamily Sector

According to the Residential Energy Consumption Survey (RECS 1990a), multifamily housing in toto—including buildings with 2–4 units as well as those with 5 or more units—accounts for 26% of the U.S. housing stock (see Figure 2-1).* There are 1 million apartment buildings having 5 or more units, with the average building having 14.4 units, for a total of 14.4 million households (Table 2-1). The American Housing Survey (AHS), which collects detailed data on the U.S. housing stock (see Figure 2-2), reports 15.2 million apartment units (AHS 1989)—800,000 more units than the RECS data—although the total number of U.S. households in the AHS agrees with the RECS data. This difference in sector size may be due to the different definitions used by the two agencies for occupied and vacant units. But by

Table 2-1

Number of U.S. Housing Unit Types in 1990 (Million Households)

Single family		64.4	68%
Detached	58.4		62%
Attached	6.0		6%
Multifamily		24.4	26%
2–4 units	10.0		11%
5+ units	14.4		15%
Mobile homes		5.2	6%
Total		**94.0**	**100%**

Source: RECS 1990a, 38.

---

* At the time of this writing, the most recently available data were from the 1989 American Housing Survey (AHS) and the 1990 Residential Energy Consumption Survey (RECS). We use present tense in describing the results from these surveys since these statistics remain a good guide to the situation in the multifamily sector today.

Figure 2-1

Total U.S. Housing Stock by Building Type

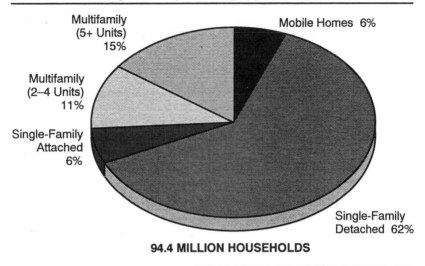

Multifamily (5+ Units) 15%

Multifamily (2–4 Units) 11%

Single-Family Attached 6%

Mobile Homes 6%

Single-Family Detached 62%

94.4 MILLION HOUSEHOLDS

Source: RECS 1990a.

Figure 2-2

U.S. Apartment Buildings by Number of Dwelling Units

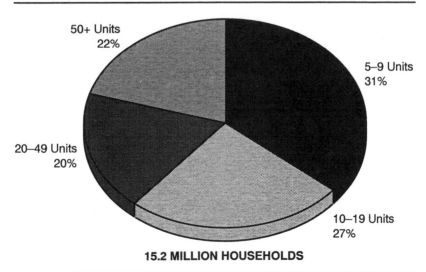

50+ Units 22%

5–9 Units 31%

20–49 Units 20%

10–19 Units 27%

15.2 MILLION HOUSEHOLDS

Source: AHS 1989.

either survey, the multifamily housing addressed here represents 15–16% of the U.S. housing stock.

Building Size

There is no typical apartment building—they range from two-story walkups to high-rise towers. Nearly a third of apartment households (31%) are in buildings with 5–9 units; 27% are in buildings with 10–19 units; 20% are in buildings with 20–49 units; and the remaining 22% are in buildings with 50 or more units (Figure 2-2).

Regional and Urban Distribution of Apartment Households

Apartment households are found in all regions of the country, in both urban and suburban locations. The Northeast and South are each home to over 4 million households, accounting for 27% and 29%, respectively, of the national total living in buildings of 5 or more units. The West and Midwest each house over 3 million households, accounting for 24% and 20%, respectively (Figure 2-3).

As a percentage of the housing stock, multifamily housing comprises 21% of the housing stock in the Northeast, 19% of the housing stock in the West, and 14% of the housing stock in the Midwest and South (Table 2-2).

Table 2-2

Regional Distribution of U.S. Housing Unit Types in 1990 (Million Households)

Housing Type	Northeast	Midwest	South	West	National
Total housing stock	19.4	22.9	32.4	19.0	93.7
Total single-family	10.0	15.5	21.4	11.3	58.2
Total multifamily 2+ units	7.4	5.5	6.7	5.4	25.1
Total multifamily 5+ units	4.1	3.1	4.4	3.6	15.2
Multifamily 2–4 units	3.3	2.4	2.3	1.8	9.9
Multifamily 5–9 units	1.1	1.0	1.6	1.1	4.7
Multifamily 10–19 units	0.7	0.8	1.5	1.0	4.1
Multifamily 20–49 units	0.9	0.6	0.7	0.9	3.0
Multifamily 50+ units	1.4	0.7	0.6	0.6	3.3

Source: AHS 1989, 34.

Figure 2-3

Regional Distribution of U.S. Apartment Households

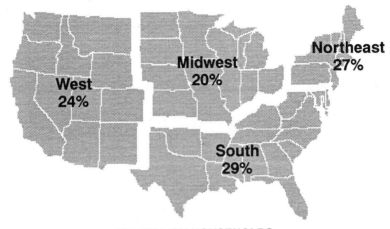

**15.2 MILLION HOUSEHOLDS
IN APARTMENT BUILDINGS NATIONWIDE**

Source: AHS 1989.

Figure 2-4

**Regional Distribution of U.S. Large (50+ Units)
Apartment Buildings**

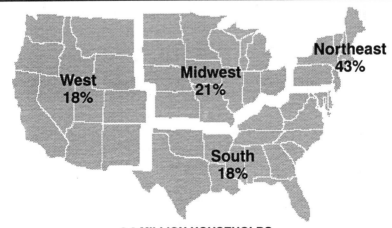

**3.3 MILLION HOUSEHOLDS
IN LARGE APARTMENT BUILDINGS NATIONWIDE**

Source: AHS 1989.

19

Table 2-3

Urban Distribution of U.S. Apartment Households in 1990 (Million Households)

Total	Central City	Suburban	Rural
14.4	7.6 (53%)	6.0 (42%)	0.8 (5%)

Source: RECS 1990a, 38.

Table 2-4

Age Distribution of U.S. Households in 1990 (Million Households)

		Year of Construction					
Sector	Total	<1939	1940s	1950s	1960s	1970s	1980s
Multifamily 5+ units	14.4	2.2	0.4	1.1	2.8	5.5	2.4
Single-family	58.4	13.8	5.4	10.9	9.4	10.5	8.3
All households	94.0	21.5	7.0	13.4	14.8	21.4	15.9

Source: RECS 1990a, 43.

The size of apartment buildings varies considerably by region, with the larger apartment buildings concentrated in the Northeast, where 42% of the buildings having 50 or more units are located (Figure 2-4). The South has a larger share of the smaller apartment buildings, both 5- to 9-unit and 10- to 19-unit buildings.

More than half of the apartment households (53%) are located in central cities, with the remainder in suburban (42%) and rural (6%) areas (Table 2-3). Large multifamily buildings account for 25% of the housing stock in central cities, 14% in suburban areas, and only 4% in rural areas (AHS 1989).

Ownership

Perhaps the single most defining characteristic of apartment buildings is that the residents are renters, not owners. Although rental households account for 30% of the total U.S. households, the majority of residents in apartment buildings rent (90%), compared with those who own their unit (10%) (Figure 2-5).

Building Age

More than half of apartment units (55%) have been built since 1970 (Table 2-4, Figure 2-6). Although we tend to think of apartment units as

Figure 2-5

Ownership Patterns of U.S. Households in Apartment Buildings

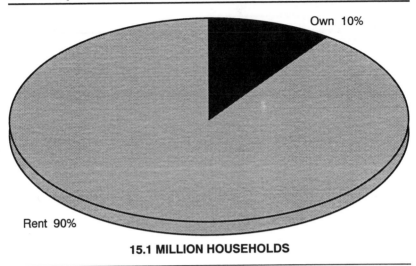

Own 10%

Rent 90%

15.1 MILLION HOUSEHOLDS

Source: AHS 1989.

Figure 2-6

Age Distribution of U.S. Apartment Buildings

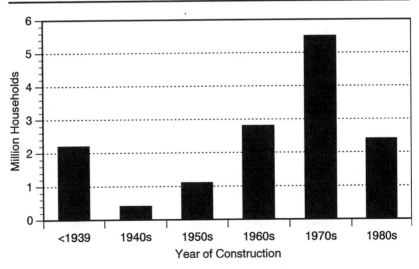

Source: RECS 1990a.

Table 2-5

**Average Household Heated Floorspace of
U.S. Building Stock in 1990**

Single-family	
Detached	1,904 ft^2
Attached	1,486 ft^2
Multifamily	
2–4 units	1,108 ft^2
5+ units	804 ft^2
Mobile home	921 ft^2

Source: RECS 1990a, 49.

being in older buildings, less than 20% were built prior to 1950. The explanation of this statistic lies in the large number of apartment buildings that were built in the South and West in the 1970s and 1980s.

Floor Area

Apartment units in large buildings (five-plus units) are much smaller on average in floor area than single family homes, having an average heated floor area of 804 square feet, less than half that of the average single family home's 1,904 square feet (Table 2-5). The median size of apartment units tends to decrease as the number of units in a building increases. Also, the size of individual units is strongly correlated with ownership: owner-occupied units are nearly twice as large as renter-occupied units.

The average heated floor area of apartment units varies by geographic location, with the larger units typically in the Northeast (1,052 square feet) and Midwest (1,008 square feet) and the smaller units in the newer buildings in the South (836 square feet) and West (807 square feet). These floor areas statistics include all apartment units, not just those in buildings with five or more units.

Household Income

Not surprisingly, residents of apartment buildings (both owners and renters) have lower incomes on average than the rest of the population. More than a quarter of the residents of large apartment buildings have total annual household incomes of less than $10,000. More than half (54%) have annual household incomes between $10,000 and $35,000. Only 8% have annual household incomes greater than $50,000 (Table 2-6,

Figure 2-7). The median income in 1989 for an apartment household was $19,100, compared with the median income of $32,100 for a single-family-dwelling household. The median monthly housing cost, the sum of all monthly housing-related payments, including mortgages or rents plus utilities, for an apartment household (both renters and owners) was $447, slightly more than the median monthly housing cost for a single

Table 2-6

Annual Income Distribution of U.S. Apartment and Single-Family Households in 1990 (Million Households)

	Total Units	<$5,000	$ 5,000–$10,000	$10,000–$15,000	$15,000–$25,000	$25,000–$35,000	$35,000–$50,000	$50,000+
Multi-family 5+ units	14.4	1.6 (11%)	2.2 (15%)	2.3 (16%)	2.9 (20%)	2.6 (18%)	1.9 (13%)	1.1 (8%)
Single-family detached	58.4	2.0 (3%)	5.6 (10%)	6.2 (11%)	9.8 (17%)	9.6 (16%)	11.5 (20%)	13.8 (24%)

Source: RECS 1990a, 55.

Figure 2-7

Household Income Distribution in U.S. Apartment Buildings

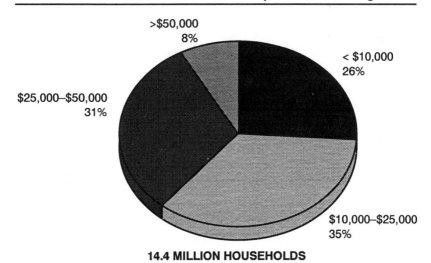

14.4 MILLION HOUSEHOLDS

Source: RECS 1990a.

23

family home, $416 (AHS 1989). Consequently, households in apartment buildings pay a higher fraction of their income for housing (27%), compared with households in single family homes (16%).

Although not all households that rent their home live in apartment buildings, the difficulties faced by many renters are applicable to a large portion of households in apartment buildings. More than two-thirds of low-income renters face serious housing problems: (1) they pay more than 30% of their income for rent; (2) they are in over-crowded quarters; or (3) they lack a full bathroom or kitchen. Nationwide, 7.6 million renter households have annual incomes of less than $10,000; however, there are only 4.4 million affordable housing units.

Similarly, although most apartment dwellers are not poor according to federal income classifications, the part of the sector that houses low-income residents warrants particular attention because housing and energy costs are such a strain on their household budgets. Nearly 4 million households in large apartment buildings (27%) are below 125% of the poverty level. Over 5 million households (37%) are eligible for federal assistance, which is defined as being below 150% of the poverty line or having less than 60% of the median state income. The fraction of income devoted to energy costs by these poorer apartment households is nearly 25%, compared with only 11% in 1970. Moreover, this fraction is three to four times the percentage of income paid by the average American family. Such a high fraction means that many low-income apartment households must make tradeoffs between paying energy bills, rent, food bills, and medical expenses (Bleviss and Gravitz 1984).

Minority and Elderly Households

Apartment households include a higher percentage of minorities than households in single-family residences. A quarter of the nation's black and Hispanic households live in apartment buildings. The 2.6 million black households account for 17% of the multifamily housing stock, compared with 9% of the single-family housing stock. The 1.7 million Hispanic households account for 11% of the multifamily stock, compared with 5% of the single-family stock. The elderly, however, are not more commonly found in multifamily housing; 19% of the households in apartment buildings are headed by someone 65 years or older, compared with 23% in the single-family stock. Nevertheless, there are 2.9 million elderly households in apartment buildings (AHS 1989).

Housing Condition

The American Housing Survey reports on the physical condition of the U.S. housing stock, based on such criteria as broken or absent

Figure 2-8

Main Heating Fuel in U.S. Apartment Buildings

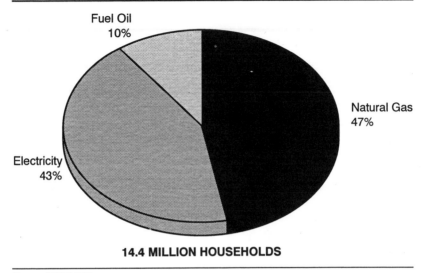

Fuel Oil
10%

Natural Gas
47%

Electricity
43%

14.4 MILLION HOUSEHOLDS

Source: RECS 1990a.

plumbing, deterioration of heating and lighting equipment, and general condition of the housing. The 1989 survey indicated that over 750,000 apartment units (5% of the occupied multifamily stock) had severe physical problems and another 660,000 apartment units (4% of stock) had moderate physical problems. These percentages are only slightly higher than for the U.S. housing stock in general.

Main Heating Fuel

Almost half (47%) of apartment households have natural gas as their primary space heating fuel (Figure 2-8). Electricity is nearly as common (43%), and fuel oil hydronic systems account for most of the remainder (10%) (RECS 1990a). Of the apartment households that heat with natural gas, roughly the same number have hydronic systems (hot-water or steam) as have forced-air central furnaces (40% and 39%, respectively). Floor or wall units account for 15%, and a variety of sources account for the remaining 6% (Table 2-7, Figure 2-9).

Over 2.5 million apartment units heat with electric central air systems (42% of the electrically heated households). A similar number (43%) heat with individual electric resistance heaters, either wall or ceiling units. A much smaller fraction of households in apartment buildings (15%) have electric heat pumps (Table 2-8, Figure 2-10).

Table 2-7

Gas Space Heating Equipment in U.S. Apartment Buildings in 1990 (Million Households)

Forced air		2.6	39%
I unit	1.6		
2+ units	0.9		
Hydronic		2.7	40%
1 unit	0.2		
2+ units	2.5		
Wall or floor units		1.0	15%
Other		0.4	6%
Total gas		**6.6**	**100%**

Source: RECS 1990a, 68.

Figure 2-9

Gas Space Heating Equipment in U.S. Apartment Buildings

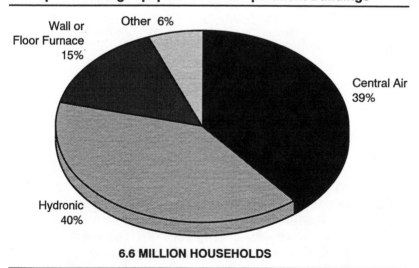

Wall or Floor Furnace 15%

Other 6%

Central Air 39%

Hydronic 40%

6.6 MILLION HOUSEHOLDS

Source: RECS 1990a.

Table 2-8

Electric Space Heating Equipment in U.S. Apartment Buildings in 1990 (Million Households)

Central air		2.6	42%
I unit	2.4		
2+ units	0.2		
Individual resistance		2.6	43%
Heat pump		0.9	15%
Total electric		**6.2**	**100%**

Source: RECS 1990a, 68.

Figure 2-10

Electric Space Heating Equipment in U.S. Apartment Buildings

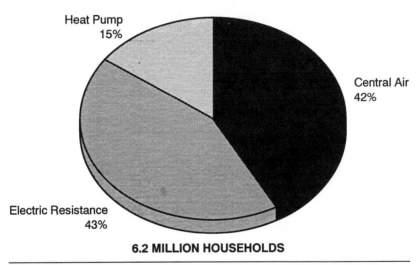

Heat Pump 15%

Central Air 42%

Electric Resistance 43%

6.2 MILLION HOUSEHOLDS

Source: RECS 1990a.

According to the RECS (1990a) data, a small number of apartment households report additional heating sources: 6% state that they use portable electric heaters, and 1% report that they use their cooking stove for space heating. Of the 700,000 apartment households with fireplaces, the majority claim to have used less than one-third of a cord of wood in the past 12 months.

Figure 2-11

Water Heating Equipment in U.S. Apartment Buildings

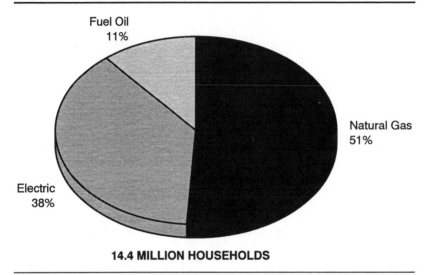

14.4 MILLION HOUSEHOLDS

Source: RECS 1990a.

Water Heating and Cooking

Roughly half (51%) of apartment households have gas water heaters, with 38% using electricity and 11% using fuel oil to provide domestic hot water (RECS 1990a) (Figure 2-11). Fifty-six percent of apartment households have hot water from central systems. Of households with individual water heaters, 71% use electricity to heat the water. Almost two-thirds (65%) of apartment units use electricity for cooking, with the remaining 34% using natural gas.

Air Conditioning

Because of the large number of new apartment buildings in the South and West, an apartment household is more likely to have an air conditioner than other households (72% saturation, compared with 67% in the single-family sector). Of the 10 million apartment households that do have air conditioning, nearly three-quarters can condition all rooms (RECS 1990a)

Appliances

Apartment households differ from other households in their ownership of major appliances (RECS 1990a). Although 100% of apartment households have refrigerators, nearly half (47%) of these are manual-defrost, compared with 20% in single-family dwellings. Apartment households are much less likely than those in single-family residences to have clothes washers (19%, compared with 93%), clothes dryers (12%, compared with 64%), and dishwashers (45%, compared with 50%).

Personal computers are less common in apartment households (11%, compared with 16% for all households), and waterbed heaters are also less common (9%, compared with 14% for all households). Saturations of smaller appliances, such as microwave ovens, televisions, and other appliances, are similar in the multifamily and single-family sectors, although appliances are typically fewer and older in apartment households.

Six percent of all apartment households report having at least one fluorescent light that is used for more than four hours per day, compared with 9% of all households.

Participation in Utility-Financed DSM Programs

Utility companies across the United States have been financing demand-side management (DSM) programs that improve the energy efficiency of the building stock. Although some utilities have targeted apartment households—particularly electrically heated units—the sector as a whole has been largely neglected. Nearly 90% of residential DSM participants live in single family or mobile homes. Apartment units account for 26% of the housing stock but only 11% of the participants in DSM programs (RECS 1990a).

Only 2% of the apartment households polled in the 1990 Residential Energy Consumption Survey (RECS 1990a) responded that they had participated in a DSM program, compared with 6% of the households in single-family houses. Of course, many residents may not have known if their building had been involved in such a program.

Conservation and Behavior

Although we have little information on how the residents of apartment buildings differ in their behavior regarding energy use, we have some indications of important differences. Apartment households differ from other households in that they tend to move more frequently, with nearly 50% of households stating they plan to move within two years (RECS 1990a). Because investments in energy efficiency accrue over several years, there may be less incentive for apartment households to invest in energy conservation.

Apartment households differ from those in single-family

residences in that they claim on average to be more likely to set back the thermostat during the day if no one is home and also to have lower temperatures at night while sleeping (RECS 1990a). Eighty percent of apartment households report that they are satisfied with their winter indoor temperature, with 13% preferring it to be warmer and 6% preferring it to be cooler, findings that are quite similar to those reported by single-family households (ibid.).

The insulation levels in apartment households are reported to be much lower than those in other households. Just over one-third (36%) of apartment households report that their units are well insulated, with another third reporting that they are adequately insulated, although the reliability of self-reported levels may not be high. The remaining third either report that their units are poorly insulated or do not know the insulation level of their dwelling. Just less than half (46%) of apartment households have storm windows, compared with 66% of other households (ibid.).

Household Energy Consumption and Expenditure

In this section we look at the amount of energy used by the average apartment household, as compared with that used by single family homes. We also look at how energy use and cost vary between different subgroups of multifamily housing, specifically how public housing and publicly assisted housing compare with the multifamily sector as a whole, both regionally and nationally.

Sectorwide Energy Consumption and Expenditure

The average apartment household uses half as much energy as the typical household in a single-family home, 51 million Btu (MBtu) per household, compared with 111 MBtu per household for the single-family sector. On a floor-area basis, however, apartment units use more energy—62,000 Btu per square foot, versus 51,000 Btu per square foot for the single family home. Public housing units use more energy than apartment units per household and per floor area but use less energy per household member—24 MBtu per person, compared with 27 MBtu per person in an apartment unit (Table 2-9, Figure 2-12).

Public housing includes a mix of federal, state, and locally owned housing, with a variety of housing types, from duplexes to high rises. Compared with how much we know about the public housing stock, we know relatively little about the physical characteristics of the publicly assisted housing stock, which is primarily private-sector housing

Table 2-9

Total Energy Consumption and Expenditure for U.S. Single-Family, Multifamily, Public, and Assisted Housing in 1990

Housing Type	Million Households	Consumption MBtu/ Household	Consumption MBtu/ Occupant	Consumption kBtu/ ft²	Expenditure $/ Household	Expenditure $/ft²
Single-family	64.4	111	39	51	1,321	0.60
Multifamily 2+ units	24.4	69	34	71	815	0.85
Multifamily 5+ units	14.4	51	27	62	677	0.84
Public housing	2.5	57	24	66	646	0.75
Assisted housing	1.7	70	25	75	863	0.93

Source: RECS 1990b, 59.

Figure 2-12

Total Energy Consumption for U.S. Single-Family, Multifamily, Public, and Assisted Housing

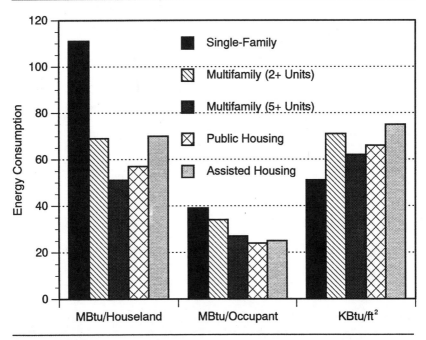

Source: RECS 1990b.

for which tenants receive rental assistance through Housing and Urban Development (HUD) Section 8 vouchers. It is likely that this stock is similar to the multifamily stock of buildings with two-plus units, as is suggested by Tables 2-9 through 2-14.

A 1986 study of energy consumption in federal public housing showed baseline consumption significantly higher than in the privately owned multifamily stock (Greely et al. 1987). In this study of 91

Table 2-10

Electricity Consumption and Expenditure for U.S. Single-Family, Multifamily, Public, and Assisted Housing in 1990

Housing Type	Million Households	Consumption MBtu/ Household	Consumption kWh/ Household	Consumption kWh/ ft²	Expenditure $/ Household	Expenditure $/kWh
Single-family	64.3	37	10,900	5.0	865	0.08
Multifamily 2+ units	24.4	20	5,800	6.0	508	0.09
Multifamily 5+ units	14.4	20	5,800	7.1	490	0.08
Public housing	2.5	19	5,700	6.6	409	0.07
Assisted housing	1.7	21	6,200	6.7	552	0.09

Source: RECS 1990b, 65.

Table 2-11

Natural Gas Consumption and Expenditure for U.S. Single-Family, Multifamily, Public, and Assisted Housing in 1990

Housing Type	Million Households	Consumption MBtu/ Household	Consumption kBtu/ ft²	Expenditure $/Household	Expenditure $/ft²
Single-family	39.5	94	42	518	5.6
Multifamily 2+ units	16.2	60	60	368	6.3
Multifamily 5+ units	8.8	41	49	246	6.2
Public housing	1.5	47	51	292	6.3
Assisted housing	1.0	70	70	406	5.9

Source: RECS 1990b, 67.

public housing projects, the median expenditure for energy was close to $1,000 per household, considerably higher than the expenditure from the RECS data (Table 2-9). This difference could be due to the different samples compared, as the RECS data set includes an additional 1.5 million state and local public housing units, whereas the Greely et al. data represent a comprehensive but not statistically representative sample of the federal public housing stock.

Apartment households consume much less electricity per household as compared with households in single-family residences—nearly 50% less on a per-household basis (Table 2-10). If we compare on a floor-area basis, apartment households consume 30% more electricity annually—7.1 kWh per square foot, compared with 5.0 kWh per square foot for single-family dwellings. Apartment household expenditure for electricity roughly follows consumption, with public housing consuming the least per household.

Gas consumption is lower on a per-household basis in apartment units but is higher when normalized by floor area—49 kBtu per square foot, compared with 42 kBtu per square foot for single-family dwellings (Table 2-11). Fuel oil consumption follows a similar pattern. Gas consumption in public housing is significantly higher per household, but only slightly higher per floor area. Consumption in assisted housing units is significantly higher both by household and by floor area.

Regional Variation in Energy Consumption and Expenditure

The regional variation in total energy consumption and expenditure is quite pronounced, with the Northeast having the highest consumption and expenditure for all fuels on both a per-household and a per-square-foot basis (Table 2-12). The West has the lowest energy expenditure of the geographic regions, although the South has slightly lower consumption on both a per-household and per-floor-area basis.

Household electricity consumption varies considerably by region, with consumption being greatest in the South, where the air conditioning demand is higher and electricity prices are generally lower. Household electricity consumption is lowest in the Northeast, which also has the highest electricity costs (Table 2-13). Household gas consumption also varies considerably by region, with consumption being greatest in the Midwest, where costs are lowest. Household gas consumption is lowest in the West, probably because of climate (Table 2-14). Apartment household fuel oil consumption is significant only in the Northeast (Table 2-15).

End-Use Energy Consumption and Expenditure

In this section we look at the amount of energy used by the average apartment household for the five primary end uses: space heating, air conditioning, domestic hot water, refrigerators, and appliances. We also analyze the regional variation in consumption and expenditure by energy end use.

Table 2-12

Total Energy Consumption and Expenditure for U.S. Apartment Households (2+ Units) by Region in 1990

		Consumption		Expenditure	
Region	Million Households	MBtu/ Household	kBtu/ ft²	$/Household	$/ft²
Northeast	6.8	91	83	1,084	0.98
Midwest	5.4	86	82	794	0.75
South	6.7	48	56	744	0.88
West	5.4	49	59	588	0.71
National	24.4	69	71	815	0.85

Source: RECS 1990c, 10, 95, 178, 281.

Table 2-13

Electricity Consumption and Expenditure for U.S. Apartment Households (2+ Units) by Region in 1990

		Consumption		Expenditure	
Region	Million Households	MBtu/ Household	kWh/ Household	$/Household	$/kWh
Northeast	6.8	14.4	4,224	498	0.12
Midwest	5.4	17.6	5,165	450	0.09
South	6.7	29.3	8,591	637	0.07
West	5.4	17.2	5,036	417	0.08
National	24.4	19.8	5,814	508	0.09

Source: RECS 1990c, 19, 104, 190, 290.

Table 2-14

Natural Gas Consumption and Expenditure for U.S. Apartment Households (2+ Units) by Region in 1990

		Consumption	Expenditure	
Region	Million Households	MBtu/Household	$/Household	$/ccf
Northeast	5.4	61	481	0.81
Midwest	4.2	87	434	0.51
South	2.6	44	252	0.58
West	4.1	42	224	0.55
National	16.2	60	368	0.63

Source: RECS 1990c, 22, 107, 194, 293.

Table 2-15

Fuel Oil Consumption and Expenditure for U.S. Apartment Households (2+ Units) by Region in 1990

		Consumption	Expenditure	
Region	Million Households	MBtu/Household	$/Household	$/Gallon
Northeast	2.9	66	471	0.97

Source: RECS 1990c, 25.

Sectorwide Consumption and Expenditure by End Use

Apartment households have different patterns of end-use energy consumption than other households because of differences in building exposure, utility metering, appliance stock, and household behavior. Space heating is typically a smaller fraction of total energy use in apartment households than in other households (Table 2-16, Figure 2-13). Appliance usage is lower in apartment households, partially because of fewer clothes washers and dryers in this stock as compared with single-family houses. Refrigerators in apartment households tend to be manual-defrost and generally smaller in size, accounting for the lower consumption in this end use. Lower air conditioning use is probably related to the smaller size of apartment units. Domestic hot water is the largest end use after space heating and represents a major target for potential energy savings.

Table 2-16

End-Use Energy Consumption for U.S. Single-Family, Multifamily, Public, and Assisted Housing in 1990 (MBtu/Household)

Housing Type	Total Consumption	Space Heating	Air Conditioning	Domestic Hot Water	Refrigerators	Appliances
Single-family	111	60	9	18	6	22
Multifamily 2+ units	68	34	5	17	4	11
Multifamily 5+ units	51	19	6	15	4	9
Public housing	57	24	4	17	4	11
Assisted housing	70	36	4	15	4	12

Source: RECS 1990b, 82.

Note: End uses do not add up to total consumption because the number of households differs by end use.

Figure 2-13

U.S. Residential End-Use Energy Consumption by Building Type

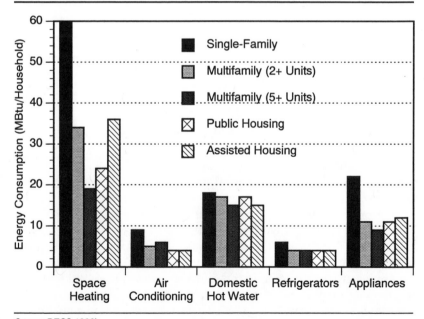

Source: RECS 1990b.

Regional variation in household energy end use probably reflects climate, age of housing stock and appliances, fuel availability, and other factors. Not surprisingly, the Northeast has the greatest consumption in space heating and domestic hot water; the South has the largest consumption for air conditioning (Table 2-17). The end-use expenditure for apartment households follows the consumption patterns in the previous table but reflects the different mix and costs of fuel and electricity (Table 2-18).

Table 2-17

End-Use Energy Consumption for U.S. Apartment Households (2+ Units) by Region in 1990 (MBtu/Household)

Region	Total Consumption	Space Heating	Air Conditioning	Domestic Hot Water	Refrigerators	Appliances
Northeast	91	54	2	21	3	12
Midwest	86	51	2	19	4	11
South	47	13	9	11	5	10
West	49	17	5	17	3	10
National	68	34	5	17	4	11

Source: RECS 1990b, 82; 1990c, 38, 123, 215, 305.

Note: End uses do not add up to total consumption because the number of households differs by end use.

Table 2-18

End-Use Energy Expenditure for U.S. Apartment Households (2+ Units) by Region in 1990 ($/Household)

Region	Total Expenditure	Space Heating	Air Conditioning	Domestic Hot Water	Refrigerators	Appliances
Northeast	1,084	434	70	178	126	319
Midwest	794	303	65	130	97	220
South	744	126	203	139	105	197
West	588	126	122	124	85	208
National	815	252	127	144	104	239

Source: RECS 1990b, 84; 1990c, 41, 126, 219, 308.

Note: End uses do not add up to total expenditure because the number of households differs by end use.

Sector Energy Consumption and Conservation Potential

From a policy perspective, it is important to know the magnitude of potential energy savings represented by the multifamily sector in order to determine what strategies are needed to achieve these savings. We begin with a characterization of the whole sector and then review the estimates of the conservation potential for these buildings.

Multifamily Sector Energy Consumption

The multifamily sector (5+ units) consumes 1.28 quads (10^{15} Btu/yr) of primary energy (accounting for electricity as generated at the power plant) or 0.73 quad of site (end-use) energy. Natural gas accounts for nearly half (49%) of the total. Public housing alone used 0.14 quad of energy (Table 2-19). The Energy Information Administration (EIA) definition of public housing includes federal, state, and municipal housing, so the sector is nearly twice as large as the federally owned public housing stock. The expenditure in 1990 for energy use in the multifamily sector was nearly $10 billion, with public housing accounting for $1.6 billion and assisted housing for another $1.5 billion (Table 2-20).

Multifamily Sector Energy Conservation Potential

The energy conservation potential in the multifamily sector—perhaps more than in any other building sector—is determined not only by the technical potential but also by the various factors that influence the likelihood of energy savings being achieved. An early estimate of the conservation potential for the entire multifamily sector (buildings with more than two units) made by the Office of Technology Assessment (OTA 1982) compared likely energy savings with possible energy savings (Table 2-21).

The assumption in Table 2-21 is that energy use in the multifamily sector will remain constant through the end of the century, in part because of the cumulative effect of demolition and addition of new stock. Of the total sector energy use (2.3 quads), OTA estimates that current cost-effective retrofit technology could save 1.0 quad (43%) a year by the year 2000. The likely savings, however, are only 0.3 quad, or 13% of the sector's energy use.

Several qualifications are in order in regarding the estimates summarized in the table because there are few statistics on the actual effects of building retrofits in apartment buildings. The data that we do have show, on average, that considerable savings are possible

Table 2-19

Energy Consumption in U.S. Multifamily, Public, and Assisted Housing Sector in 1990 (Quadrillion Btu/Year [Site Energy])

Sector	Total Consumption	Electricity	Natural Gas	Fuel Oil
Multifamily 5+ units	0.73	0.28	0.36	0.08
Public housing	0.14	0.05	0.07	—
Assisted housing	0.12	0.04	0.07	0.01

Source: RECS 1990b, 61.

Table 2-20

Energy Expenditure in U.S. Multifamily, Public, and Assisted Housing Sector in 1990 ($Billion)

Sector	Total Expenditure	Electricity	Natural Gas	Fuel Oil
Multifamily 5+ units	9.8	7.0	2.0	0.5
Public housing	1.6	1.0	0.4	0.1
Assisted housing	1.5	1.0	0.4	0.1

Source: RECS 1990b, 63.

Table 2-21

Energy Conservation Potential in U.S. Multifamily Housing in the Year 2000 (Quadrillion Btu/Year [Primary Energy])

Multifamily Building Type	Trend Energy Use	Technical Savings Potential	Likely Savings
Low-income	0.6	0.2	0.1
Moderate- and upper-income			
Master-metered	0.9	0.4	0.1
Tenant-metered	0.8	0.4	0.1
Total	**2.3**	**1.0**	**0.3**

Source: OTA 1982, 35.

from low- and moderate-cost retrofits, but that savings achieved in individual buildings may vary considerably. Because each structure is a unique combination of design, siting, construction, and previous retrofits—and because the behavior of occupants, owners, and weather is often unpredictable—the actual performance of retrofits may vary substantially from that predicted.

A second study to estimate the potential energy savings in the multifamily sector was conducted by researchers at Lawrence Berkeley Laboratory in 1988 (Goldman et al. 1988c). Using the results from their database of measured retrofit costs and savings, Goldman and his colleagues extrapolated the savings to the U.S. multifamily stock, using information on the building characteristics from the 1984 Residential Energy Consumption Survey (RECS 1984).

After adjusting for differences in climate and initial pre-retrofit consumption, Goldman et al. found that typical retrofits could save about 0.2 quad per year (in primary energy) and that intensive retrofits could save about 0.5 quad per year. This estimate, representing 10–22% savings from retrofits, is well below the technical potential for conservation of 43%, as estimated by the Office of Technology Assessment. Using actual costs for buildings in the database, they estimated that the cost of retrofitting the entire multifamily stock with "typical" retrofit packages would be between $7.5 and $11 billion, and for the more intensive retrofit packages, the cost would be between $27 and $32 billion (Goldman et al. 1988c).

Although a large potential for energy savings in multifamily buildings does exist, achieving these savings is probably more challenging in the multifamily sector than any other, given the limitations it faces in availability of capital, retrofit information, and public programs, as we discuss in the following sections on building owners and barriers to multifamily retrofits.

Apartment Building Ownership and Trends

Apartment buildings are owned by individuals, partnerships, and housing management corporations as well as by government agencies and institutional investors, such as companies and pension funds. Individuals mostly own smaller properties but also account for 20% of buildings with 50 or more units (Goodman and Gurpe 1995). The past decade has seen a sharp decline in rental property ownership by general and limited partnerships, which accounted for only 3% of ownership in 1991. Institutional investors tend to hold properties for six years or longer and are likely to have a higher-than-average interest in capital improvements with relatively long paybacks (Table 2-22).

Apartment building owners can be classified in two general categories—those who own buildings occupied by moderate- to high-income residents and those who own buildings occupied by low-income residents. The distinction made between these two groups is due in part to the different economic constraints associated with the two income levels (OTA 1982).

Owners of Moderate- to High-Income Housing

This subsector includes owners of apartments, condominiums, and cooperatives. Ownership types include:

- Individuals

- Members of condos and cooperatives

- Local partnerships

- Syndicated partnerships

- Corporations—developers, insurance companies, pension funds

---

Table 2-22

Retrofit Payback Criteria, Holding Periods, and Access to Financing and Advice for Different Types of U.S. Apartment Building Owners

Building Owner Type	Typical Payback Criteria	Building for Own Use	Expected Holding Period	Access to Capital	In-house Professional Advice
Owner-occupants					
Large corporations	3–5 years	Yes	Long	Good	Good
Small businesses	1 year	Yes	Long	Poor	Poor
Owner-occupants	1–3 years	Yes	Long	Poor	Poor
Investor-owners					
Institutional owners	5–7 years	No	Long	Good	Good
Development companies	1–3 years	No	Short	Fair	Good
Partnership syndicates	3 years	No	Short	Fair	Good
Local partnerships	1–2 years	No	Short	Poor	Fair
Individuals	1 year	No	Mixed	Poor	Poor

Source: OTA 1982, 11.

The ownership trend is toward investment partnerships, especially large syndications, which are now the most common form of ownership. Partnerships can support energy retrofit activities, but usually only at the time of syndication or refinancing. The short period of time for which these partnerships hold the buildings (five to seven years), however, means that economic payback on investments must be rapid or that sale price must increase as a result of the retrofit (OTA 1982).

This sector is also stratified between large-building owners and smaller-building owners. Corporations and syndicates tend to own larger buildings and have a strong edge in technical expertise and financial resources. Smaller-building owners tend to be individuals with limited access to both of these.

One of the first energy retrofits sought by owners in this subsector is to install individual-unit submeters in centrally supplied buildings to remove energy costs from their direct responsibility. The trend toward tenant metering (used in almost half of all units) means that the incentives for owner investment in energy conservation retrofit are shrinking.

There is also a trend in this subsector toward the hiring of professional property managers (currently used in 30,000 to 40,000 apartment buildings). In professionally managed buildings, building managers are likely to be the key to implementing low-cost/no-cost measures, as well as to ensuring good operations and maintenance practices. Building managers can also influence owners to make cost-effective capital retrofit measures.

The most influential organizations of building owners in this subsector appear to be the trade associations, such as the Institute for Real Estate Management (IREM), with 30,000 to 40,000 members, who manage about one-third of privately owned apartment buildings.

Owners of Low-Income Housing

The types of low-income housing ownership are equally diverse, falling into four general categories:

- *Established owner-manager:* This landlord purchases property as a long-term investment for financial security.

- *Trader:* The trader purchases buildings with the idea of making a profitable and often quick resale.

- *Operator:* This landlord often cuts back on maintenance to increase profits. Although the building deteriorates, the rents usually remain stable.

- *Rehabilitator:* This landlord upgrades deteriorated buildings and increases rents as the building improves.

Of these four, the established owner-manager is the most likely to engage in energy retrofits without displacing low-income residents afterward; this ownership category appears to be the most representative of low-income multifamily housing.

There are virtually no detailed data on apartment building ownership. Most of what is known about ownership of buildings is known from real estate trade literature and the expertise of real estate analysts and operators. In some states, ownership is hidden by various devices permissible under state laws.

Public and Publicly Assisted Housing

According to RECS (1990b), various levels of government own over 2.5 million units of public housing. The federal government owns 1.25 million of these units. In addition, the federal government assists 1.75 million units of privately owned housing, largely subsidized (Section 8) housing, but including other HUD programs. HUD spends approximately $3 billion each year to pay all or part of the energy bills for households in public and assisted housing. In addition to federally assisted housing, some states (such as New York and Massachusetts) have state-assisted housing.

Like the multifamily sector in general, publicly assisted housing is a mix of low- and high-rise buildings; centrally and individually metered systems; and new and old building types. Although the physical characteristics of the public and assisted housing stock do not differ greatly from those of the rest of the multifamily stock, different programs and financing are often needed for improving their energy efficiency, as discussed in Chapters 4 and 5.

Will Apartment Building Owners Invest in Energy Efficiency?

Given the different types of multifamily building owners, are there any patterns as to which building owners are likely to invest in retrofit activity? The OTA study lists economic criteria that different building owners hold in deciding whether to undertake energy efficiency improvements to their properties (OTA 1982).

Owner-occupants of apartment buildings expect to hold their buildings for a long time and would benefit from retrofit, but they are severely constrained by lack of access to capital and generally cannot tolerate losses in cash flow. On the other hand, large apartment properties owned by institutional organizations, such as insurance companies and pension plans, routinely make capital investments in energy efficiency if such investments will pay back in less than five to seven years.

Trends in Multifamily Stock

Several trends are significant in characterizing the multifamily stock over the past decades. Changes in the total number of units, vacancy rates, and the mix of low- versus high-rent units have all had an impact on energy use in apartment households.

New apartment construction reached a high mark in late 1985 and early 1986, averaging an annual rate of about 700,000 units. This peak followed four strong years of apartment construction that represented a bounce-back from the depressed levels of 1981–1982. The vigorous activity during these years was spurred by the tax provisions of the Economic Recovery Tax Act of 1981 (ERTA), which allowed for accelerated depreciation, short tax life, high marginal income tax rates, and low capital gains tax rates that were extremely favorable to real estate (Apgar et al. 1991).

The recent declines in apartment construction are not only the consequence of the overbuilding of the 1980s (made more severe by the slowdown in the U.S. economy), but are also a result of the much less attractive tax incentives for apartment construction in the Tax Reform Act of 1986 and the credit crunch caused by the Thrift Reform Act of 1989. There seems to be little stimulus left to construct low-rent apartment units without further tax incentives or subsidies (Apgar et al. 1991). Apartment housing starts in 1990 were 310,000 units, less than half the number of five years previous.

In 1991, total housing starts stood at their lowest level in 40 years. Apartment starts were particularly hard hit, down 74% from their previous peak—174,000 units, compared with the high of 1,048,000 units in 1972 (JCHS 1992). (In 1991, there were as many new mobile homes as multifamily starts.) Only a handful of states registered increases in apartment building permits in the early 1990s. And even in these states (Montana, Wyoming, Utah, Texas, and Arkansas), the number of permits was generally small (ibid.).

During the 1980s, the number of apartments renting for less than $300 per month (constant 1989 dollars) dropped by nearly 1.8 million units; during the same period, the number of units with rents above $500 more than doubled. The loss of low-cost apartments is reflected in the increase in median rents. Since 1981, rents have increased 16% faster than prices in general and now stand at their highest levels in two decades (Apgar et al. 1991).

Vacancy rates for apartment units climbed throughout the 1980s. The highest vacancy rates are found in the South, some 30% above the national average. Much of the high vacancy status resulted from the surge in multifamily construction between 1983 and 1987, which ran ahead of demand in several large markets. Overbuilding in such areas as Dallas and Houston has grossly distorted the trends for entire regions (Apgar et al. 1991).

A report by the National Low-Income Housing Preservation Commission (1988) predicts that some 523,000 units subsidized by HUD are at risk of being lost over the next 15 years, either because of expiring use restrictions (which permit conversion to higher-income occupancy) or because of weak financial conditions that result in disinvestment and abandonment. We will discuss in Chapter 4 how some of these at-risk properties were able to remain as affordable housing through efforts to lower the cost of energy bills.

Rent burdens increased in the 1980s. The gross rent burden (payment to landlord plus fuel and utilities as percentage of household income) rose above 30% in 1991, for only the third time in 25 years. Contract rent burden (payment to landlord as percentage of household income) reached a record high of 26% in the early 1990s (JCHS 1992). The majority of unassisted poor renters devote at least half of their income to housing, and even paying this large a share is no guarantee that they can live in decent housing (ibid.).

Affordable housing has been in long-term decline, where *affordable* refers to subsidized units (which are affordable by definition) plus unsubsidized units with monthly gross rents of less than $300 measured in 1989 dollars, or roughly 30% of the monthly poverty-level income for a family of four.

Over the past 15 years, the number of market rate units renting for $300/month or more has steadily grown, whereas the supply of subsidized or otherwise affordable housing has steadily shrunk, primarily because of the continuing loss of unsubsidized low-cost units. Although the number of subsidized units nearly doubled between 1970 and 1990, from 2 million to 4 million, the increase was not enough to offset the losses of the low-cost unsubsidized units. The total number of affordable units dropped from nearly 10 million in 1974 to 9 million in 1990 (JCHS 1992).

The multifamily sector continues to face both short- and long-term obstacles. Most immediately, the combination of overbuilding in many areas, the loss of investment incentives under tax reform, and the crisis in the banking industry has brought apartment construction to a virtual standstill. Over the longer term, demographic forces—particularly the movement of the baby-boom generation into the prime homeownership years—might shift even more demand from the multifamily to the single-family market.

The 1990s are likely to see apartment construction remaining low, allowing current and future vacancies to be slowly absorbed (Apgar et al. 1991). Multifamily housing will continue to be lost because of abandonment. As a result of a combination of upgrading and demolition, the unsubsidized multifamily stock is disappearing from the market at a rate faster than subsidized multifamily stock is being added. Despite the low level of multifamily housing starts, production of market-rate rental units is expected to rebound, but only slowly, as vacancies are absorbed and markets tighten. These changing demographic and market forces will do little to help the housing conditions of the poor. Along with economic recovery will come rising rents, adding pressure on poor households that are already spending large shares of their income on housing. Without expanded public subsidies, many households will be excluded from sharing the prosperity brought by resumed economic growth (JCHS 1992).

Conclusion

Although this chapter has presented a multitude of numbers, the multifamily sector is more than a statistical compilation of floor areas, numbers of apartments, and units of energy consumption. This sector represents the homes of countless individuals, many of them poor, for whom the tradeoffs between better housing, food, and fuel are pressing. The potential for saving energy and improving the quality of life is great, and the barriers to furthering energy efficiency in multifamily housing are indeed real, but they are not insurmountable. Much progress has been made in finding ways to break through these obstacles. The following chapters draw on what has been learned from past and recent efforts to improve energy efficiency in the multifamily sector. We examine the technical opportunities for saving energy, the programmatic means for achieving savings, and the ways in which efficiency improvements can be financed.

Energy Conservation Opportunities

Optimizing the energy efficiency of an apartment building ultimately rests not only on changes in the structure and equipment of the building, but also on the actions of management, staff, and residents. Although energy conservation measures can improve efficiency of mechanical systems (heating plants, distribution systems, controls, and other heating, ventilation, and air conditioning [HVAC] equipment), the building shell (windows, insulation, air sealing, and so forth), and other building systems (domestic hot water, lighting, and appliances), the proper maintenance and operation of all the systems are as important as physical improvements. Developing an understanding of equipment and energy costs, training for operation and maintenance, establishing accountability, and providing the right balance of human versus automatic control are ways to effect the behavioral changes needed for the physical changes to result in reliable energy savings.

Much technical progress has been made in apartment building energy conservation in the past two decades. Following the 1973–1974 oil crisis, which sparked closer scrutiny of energy use in buildings, early measurements and engineering-based analyses identified the potential for large energy savings in apartment buildings (Beyea et al. 1977). However, lack of reliable information on how to improve

Loretta Smith, formerly Research Associate at ACEEE, made substantial contributions to this chapter, particularly the material on domestic hot-water systems, lighting, and appliances.

energy efficiency was identified as a major barrier to effective conservation in the early 1980s (Bleviss and Gravitz 1984). By the end of the decade, technical know-how had become less of a limitation. In summarizing the Multifamily Building Technology panel of the 1988 ACEEE Summer Study, Hewett (1988b) highlighted the progress made during the 1980s in testing retrofits, developing audit tools, and analyzing energy use in apartment buildings. Hewett et al. (1994) review how empirical work has revealed sets of retrofit measures that have reliably improved the efficiency and performance of certain types of apartment building systems. A recent special issue of *Home Energy* magazine features a collection of articles by leading multifamily sector conservation experts, providing up-to-date information techniques for improving efficiency in a variety of building types (*Home Energy* 1995; also see Appendix B).

Thus, practitioners can now learn "what works" for some important building types in several regions of the country. For example, we now have fairly extensive information on improving energy efficiency of fuel- and electricity-heated buildings in heating-dominated regions of the country. However, space heating accounts for less than half of apartment building energy use nationwide. Furthermore, conservation knowledge has yet to be widely applied, and gaps remain in many areas. Additional testing is needed of retrofits that show promise, and we need more data as well as better analytic tools for treating the many complexities of energy use in apartment buildings.

Goldman et al. (1988a, 1988b) reported on the costs and achieved savings from retrofits for 191 apartment buildings containing a total of over 25,000 dwelling units. This survey was drawn from the experience of several leading energy conservation centers around the country. Goldman et al. reported median savings of roughly 15% of pre-retrofit energy use, but with substantial variation: savings were between 10% and 30% for three-fifths of the buildings surveyed. There were many outliers, including a few buildings in which energy use increased after retrofit. High per-unit pre-retrofit consumption was a significant predictor of savings, confirming the value of this factor as a screening tool for retrofit programs.

Although it is convenient to discuss technical retrofits separately for major physical components of a building—as we do in this chapter—it is essential to keep in mind the integrated nature of the "building as a system." Substantial, persistent, and cost-effective energy savings are rarely obtained with a piecemeal approach to retrofit. For this reason, auditing and planning are the necessary first steps in the retrofit process. It is at this stage that critical decisions must be made, based on an understanding of the building, its context, and the available resources. The technical opportunities for improving the efficiency of mechanical systems, shell components, and other energy-using equipment in apartment buildings can then be explored within an overall plan for the building as a system.

The behavioral and physical aspects of apartment building energy use are intimately linked. Therefore, rather than treating behavioral issues separately, we address them as part of retrofit planning and as part of the implementation of various physical retrofits. Behavioral research on apartment building energy conservation has shown that engaging the people affected by a retrofit—residents, staff, and management— provides opportunities for success. Conversely, disregarding human needs and constraints is a recipe for failure.

Audits and Retrofit Planning

Auditing refers to a building evaluation and diagnostic process that leads to decisions about retrofit actions to be taken and measures to be installed. The process of auditing an apartment building for energy retrofits varies with the type of energy conservation program and the resources available.

The comprehensiveness of an audit depends on the program's objectives. Audits provided in response to state or utility program requirements or as part of marketing efforts that seek widespread participation tend to be less comprehensive. Audits performed as part of pilot studies or demonstration programs or in conjunction with research programs are more comprehensive and often the most detailed in terms of data gathering. Audits done in preparation for performance contracting or shared-savings arrangements, in which there is a financial stake in the reliability of estimates, must thoroughly uncover the key determinants of energy consumption. But audit detail is not equivalent to audit effectiveness. Success depends on the auditor's developing a sound understanding of the building as a system, including both physical and behavioral factors, even for limited retrofit programs seeking to install only a few relatively low-cost measures.

In this section, we discuss types of audits, starting with comprehensive approaches and then turning to approaches that are more limited in resources. An important notion is that of the auditor as project manager—the individual who understands the big picture in terms of both opportunities and resources and who provides the follow-through needed to ensure success. We also review audit validation results in the multifamily sector. Further details on diagnostic tools, techniques, and computerized audits applicable to apartment building retrofit planning are given in Appendix C.

Comprehensive Audits

No doubt many buildings are in such poor condition that they present obvious component-specific problems that can be fixed with immediate benefits. The need for such major rehabilitation is an important opportunity for making energy-related improvements. In other cases, specific opportunities may be available, such as common-area lighting efficiency improvements or water conservation measures. Generally, however, it is necessary to analyze how the building as a whole is functioning and how it will change with the retrofit options. A number of challenging questions must be addressed. What type of property (private, public, or subsidized) is involved? Who pays the bills? What is the recent consumption pattern, and how does it compare with that of similar buildings in the region? What are the conditions in the dwelling units and common areas? What resources can be made available, by the owner or other sources, for retrofit? Answers to these and other questions are needed for an auditor to see how the various aspects of the building interact to create an energy-wasteful situation. Developing an understanding of the building as a system—including the perceptions and behavior of the residents, staff, and management along with analyses of the structure and its components—is crucial for successful retrofit planning.

The building-as-a-system approach is particularly important because of the numerous interactions among the various elements of a building that can be affected by retrofit. Shell tightening can affect air quality and heat distribution. Insulation can affect air leakage characteristics and cause unforeseen moisture problems if inappropriately applied. Shell measures that reduce heating load can result in lower seasonal heating system efficiency, reducing some of the potential energy savings. Electrical and auxiliary equipment improvements can increase a building's heating load but reduce its cooling load. Window retrofits without attention to heating system control may lead to tenants' opening windows for comfort control, resulting in zero or negative energy savings. New thermostats may be ineffective in improving system control unless residents and building staff are educated about how they work and how to use them.

In general, the steps in an energy conservation audit include diagnosis of building energy use, identification of potential retrofits, and financial analysis for the retrofit project. Diagnosis involves the systematic collection and analysis of extensive information about the building, its physical characteristics, and its energy use history (including prior retrofits or other changes), as well as interviews with management, staff, and residents (Dutt and Harrje 1989). It is often valuable to supplement this information gathering with on-site measurements related to specific building characteristics and equipment (such as boiler flue gas analysis). A number of state and local energy conservation programs have prepared apartment building audit manuals that can be useful in planning the diagnosis stage of an audit. Diagnostic tasks can also be facilitated by computerized audit tools that help identify the information that must be gathered (checklists), organize the information, provide reasonable assumptions to fill in gaps, and help guide cost/benefit analysis of recommended retrofits.

There appears to be no substitute for past experience in guiding the audit process. The Environment and Energy Resource Center (EERC) in St. Paul, Minnesota, has one of the leading programs in apartment building energy conservation. When asked about their preferred method, Dave Lido of EERC responded, "We really don't have any. We know from past experience what kind of savings we can expect from certain retrofits on different types of buildings, so we can simply apply these percentages to the present consumption data and get some good predictions of the savings we are likely to see. The secret is sticking with the project, making sure things are done right and everything is functioning the way it should. We have the building owners let us take over the operation of their buildings, so we make sure things work." Such glibness about lack of a preferred method should not mask the fact that seasoned practitioners draw on a variety of often-sophisticated methods. The skill of groups like the EERC comes from experience: in billing data analysis; in thorough field inspections, including appropriate on-site tests and measurements, computerized audits, and retrofit monitoring; and in careful follow-up, dating from past work such as that reported by the then–Energy Resource Center (Robinson et al. 1986).

John Snell of Boston's Citizens Conservation Corporation (CCC) traced an evolution in CCC's experience in apartment building audits and retrofit planning. In early efforts, CCC drew on measure-by-measure projections of energy savings, often provided by different consultants and usually constrained by the need for short paybacks of less than five years. Taken together, such packages of discrete conservation measures resulted in overoptimistic expectations, and the program struggled to achieve cost-effective savings. CCC was also inexperienced with how various measures worked in practice for the stock of buildings being addressed. Overestimation of savings was common when relying largely on engineering estimates of savings and payback times without a substantial base of measured savings results. Although such experience led to greater caution, it also led to a realization that an approach largely built around discrete, supposedly short-payback conservation measures was too limiting to address the energy consumption problems of the buildings CCC was seeking to retrofit. CCC then worked to extend the payback timeframe to ten years, which allowed inclusion of extensive building refurbishing and system replacement measures and began taking a comprehensive major retrofit approach. The results of such "holistic" efforts (see Case Study: A Holistic Approach to Audit and Retrofit, below) proved that very substantial savings could be achieved. However, one cannot extrapolate the specific savings expectations from this approach to other building stocks in different regions with different institutional settings.

Given such experience, CCC's audit projections are now much closer to achieved energy savings and cost-effectiveness. The resulting retrofit plans must result in payback periods no longer than the duration of the performance contract, typically on the order of ten years. This allows much more than low-cost, short-payback measures but avoids isolated use of long-payback measures, such as window replacements. Long- and short-payback measures are combined with monitoring, education, and follow-up to yield an attractive overall package having a high probability of producing substantial savings that closely match the audit estimates. CCC's Lillian Kamalay notes, "If you have the people (residents, staff, and management) on your side, then you have more than just the physical measures to rely on" for achieving the estimated savings.

Prescriptive Approaches

The comprehensive audit approach just described can be thought of as an ideal that can be reached given adequate skills and resources. Many programs seek to accomplish much less on a per-building basis but can reach many more buildings with a limited, but potentially effective, set of measures. Such prescriptive approaches to audit and retrofit are often used in government and utility conservation programs that are charged with covering a substantial portion of the building stock in a given area.

In utility demand-side management (DSM) programs, audits usually follow a checklist approach based on prior calculations of the impact of various retrofit options. Options may be scored according to impacts on both consumption (kWh) and utility peak demand (kW). For instance, replacing all refrigerators over five years old and replacing common-area incandescent lighting with more efficient products may have an average net positive effect on peak demand in a particular geographical area. A DSM program would be set up to install this equipment, and the audit procedure would simply inventory all such replaceable units. In any particular building, other retrofit opportunities may exist that are cost-effective for the owner but out of bounds for the DSM program because they fail the utility's avoided costs tests—a very different standard than energy savings paybacks.

Code enforcement inspections (as well as most home energy rating systems) typically use a checklist approach. Wisconsin, for instance, has a rental energy code that must be met whenever a building changes ownership. It is strictly a prescriptive measures code, however, so inspectors are simply asked to confirm that doors have weather-stripping (whether they leak or not), windows have storms, and attics have insulation (if they are reasonably accessible). The inspection documentation is a "Certificate of Compliance" and simply records whether each measure passes, fails, is not applicable, or is not accessible.

Although the more sophisticated low-income weatherization programs take a holistic approach, less experienced programs often focus solely on apartment-level retrofit options. Such programs may ignore overall building conditions and central-heating-system retrofit opportunities, even when most of the building's tenants meet program low-income guidelines. There may be little or no opportunity to address climate-control systems in the individual apartments. With limited whole-building multifamily-housing experience, auditors are likely to rely on a fairly narrow set of architectural measures taken from their single-family-housing experience. The result is often either a windows program or no-cost/low-cost retrofits that generally result in only limited savings.

Prescriptive, checklist-based approaches to apartment building audit gave rise to a number of handbooks by the early 1980s, which themselves became precursors to computerized audit tools. Handbook-based approaches are often useful in apartments with individual heating systems. Anomalies in individual units can be identified by examining and comparing the pre-retrofit energy consumption of the various apartments. As is the case for attached single-family units, the variation in energy use among identical apartments is likely to be wide. Billing data can be used to identify high-consumption units for further audit and for focused efforts in physical retrofits and tenant education. Detailed audits on each apartment in a building with 20 nearly identical units may be unnecessary.

Alternatively, in larger, individually metered buildings, it may be prudent to perform a prototype trial on one apartment. The retrofitter can select a "typical" apartment with average consumption and install a best-guess analysis of both architectural and mechanical retrofits. After short-term monitoring of the energy savings, the retrofit procedure can be modified to improve performance. Once the proto-type retrofit meets expectations, the process can be duplicated in the remaining units. Although this process may be expensive and time-consuming, it will be far less expensive than making the same mistake 50 times in an entire complex before discovering that there were better options.

In practice, of course, the auditor need not run all calculations on every building. Rather, past experience and a knowledge of possibilities with the present building drive the selection of measures. This type of knowledge has been built into a "smart" computerized checklist developed by the Vermont Energy Investment Corporation (VEIC) for use in DSM programs (Hamilton et al. 1994). The VEIC "Smart Protocol" uses a series of look-up sheets based on common variables for determining the cost-effectiveness of a variety of retrofits. In a lighting program, for instance, the auditor considers such variables as existing incandescent wattage, daily burn time, the characteristics of replacement fixtures, and the local avoided electricity cost to select the most cost-effective retrofit. In any case, documented benefit/cost calculations help verify the recommendations and serve to justify the retrofit plans to building owners and program administrators. Further details on audit tools applicable to apartment buildings are provided in Appendix C.

The Auditor as Project Manager

Despite the availability of audit software, guides, and paperwork, the auditing task is only a small part of the overall process of carrying out a retrofit from inception through construction and on to follow-up monitoring. In fact, in many instances, the title of the key individual performing these tasks is not "auditor" or "inspector," but rather "project manager." Experience has shown that only through comprehensive and detailed tracking of every element in the process can so complex a system function as intended. The project manager must work closely with building owners, managers or supervisors, representatives of funding sources, code officials, maintenance personnel, contractors, and tenants alike. Having a point person in this role is critical, for example, in helping private building owners develop the confidence to follow through with the audit recommendations. It is the role of the project manager not only to identify energy conservation opportunities, but also to make a careful assessment regarding costs, likely benefits, interactions of measures, resources for getting the job done, and what a building owner desires and is able to agree to.

Thus, identifying and planning apartment building retrofits is a job for experienced professionals. The project manager must keep the big picture in mind and know which specialists (such as HVAC or plumbing and heating contractors) are needed to handle specific tasks, and also what their capabilities and limitations are. A number of firms and organizations specializing in energy conservation and that have developed expertise in carrying out retrofits of apartment buildings can help identify those who are capable of providing this broad technical service. Some are also capable of training others to develop this level of expertise. Appendix B lists a number of organizations that can either directly provide, or offer guidance about how to obtain, these services.

Audit Validation and Performance

The validity of energy audits can be evaluated by comparing predicted with measured energy savings. Goldman et al. (1988b) summarized comparisons between predicted and measured savings for six apartment building retrofit programs covering 54 projects. Median measured savings exceeded audit predictions in four of the six programs. In 11 projects of the Energy Resource Center (ERC) of St. Paul, Minnesota, discussed by Goldman et al. (1988b), median measured savings were 27% higher than the audit predictions but exhibited less variation than those of other programs. ERC carefully monitored energy use and operations of the retrofit buildings, which contributed to the good performance. The ERC efforts were for shared-savings projects, which tend to employ conservative approaches (underestimating savings) for project financial planning rather than striving for a mean error of zero in actual versus audit-predicted savings (the "disinterested," scientific approach). On the other hand, CCC's experience suggests that for well-targeted, comprehensive retrofits, accuracy rather than conservatism is desirable. Downward-biased savings estimates may fail to justify the major investments (for example, major shell upgrades or mechanical system replacements) needed to obtain deep cuts in energy consumption.

From the efforts reviewed by Goldman et al. (1988a, 1988b) and the CCC experience, we can highlight the elements of successful apartment building audits:

- Analysis of utility bills and screening buildings to target those with major retrofit needs
- Careful engineering calculations of monthly loads, calibrating the estimated building and equipment parameters to measured pre-retrofit consumption

- Thorough site inspection, including the condition of dwelling units as well as mechanical equipment and operating procedures

- Interviews with management, staff, and residents to develop an understanding of actual operations and the concerns of the different parties

- Reliance on experienced technical personnel and energy conservation experts in developing savings estimates for the planned retrofits

Achieving a good match between achieved savings and audit estimates also depends on careful quality control during the retrofit process and follow-up visits after installation. It is appropriate for most programs to focus resources on the most energy-inefficient buildings identified. Given a thorough audit, appropriate targeting of buildings, and adequate resources, it is likely that one can more reliably attain substantial energy cost savings (for example, on the order of 25–50% or more of pre-retrofit bills with paybacks of under ten years) by using a comprehensive package, as opposed to the modest savings (for example, on the order of 10%) achieved using a more limited approach.

CASE STUDY
A Holistic Approach to Audit and Retrofit

Citizens Conservation Corporation (CCC) of Boston, Massachusetts, has evolved a comprehensive approach to auditing and retrofitting apartment buildings that involves both the technical and behavioral aspects of energy use. We dub this approach "holistic" and note that, coincidentally, SyrESCO has developed a similar approach, which it also terms "holistic" (SyrESCO 1994).

Conversations with CCC project managers Lillian Kamalay and John Snell* have provided many useful insights, which we synthesize here along with lessons reported by other practitioners and researchers. High energy use comes not only from problems with "hard" components—the structure and physical plant—but also from the behavior of the "soft" components—management, staff, and tenants. An engineer might view high consumption in a building as a result of poor optimization. However, an energy-wasteful situation might actually be well optimized from the perspectives of the individuals involved and given their institutional arrangements (DeCicco and Kempton 1986).

---

* Much of the discussion in this case study is based on phone interviews conducted by John DeCicco with the cited CCC staff.

This insight is important because optimized conditions tend to resist change unless some of the fundamental parameters are changed.

An essential part of any audit, therefore, is developing an understanding of the human as well as the physical components of the building as a system. This part of the process entails interviewing all of the relevant players—management, staff, and tenants. The auditor needs to develop a relationship with key players. Most apartment buildings have someone who is the effective site manager. In some privately owned buildings, this may in fact be the owner. In other cases, it may be someone in a management firm or housing authority, depending on the specific ownership/management situation. The auditor must also learn about other people involved. Who pays the bills? Does anyone track them? What are tenants' views about comfort and building services? The auditor needs to cross-check and compare the information obtained from management, staff, and tenants, who have different views, no one of which will tell the whole story.

The fundamental role that behavioral aspects play in apartment building energy conservation is illustrated by a recent effort by the Chicago Housing Authority (CHA) and CCC. As recounted by CCC's Lillian Kamalay, CHA had been collecting utility bills from all its projects and aggregating them by fuel type. The result was that those who looked at utility costs never saw what was going on in individual buildings. Part of the contractual arrangement between CCC and CHA was for the Housing Authority to provide monthly utility billing data for the past three years. CHA initially could not meet this request with its existing accounting system; after some months had passed, the Housing Authority hired a new person specifically to track utility bills. By going through the process of gathering information needed for energy bill analysis, CHA management learned about how and why to track consumption. Thus, by the time it was through with the audit, CCC had already effected a change in the way CHA accounted for energy use. This case may be extreme in that it resulted in a new person being employed, but it was appropriate for an authority as large as Chicago's. Even in smaller projects, important changes in understanding and behavior are accomplished by putting building management through the rigorous audit process needed to assess energy consumption and to plan a retrofit. Securing a commitment by owners or management to supply energy consumption records and go over them with the auditor should be a part of standard apartment building retrofit planning procedures.

Also during the course of the Chicago effort, CCC directly interviewed management and maintenance staff regarding conditions in a large housing project, complementing the physical inspections carried out by the technical auditors. However, the apartments were far too numerous for CCC auditors to talk to a significant number of residents. Therefore, a local minority subcontractor was hired to train some residents to interview a broader sample of residents throughout the project, allowing survey coverage of 5% of the occupied units. Particularly in projects of this scale, such resident surveys are essential for verifying the systems diagnosis and retrofit plans developed through the technical audit. The survey also helped build contacts with residents. Resident interviews provide vital information about potential problem areas with building services (as well as with the responsiveness of maintenance staff) for central-heating and hot-water systems in a large project. The extra effort expended to conduct a survey highlights the importance CCC places on resident-provided information for developing and validating a workable plan.

Once a "hard" retrofit plan is developed, implementation must include attention to the "soft" aspects of the situation. Kamalay captures this concept in three imperatives: monitoring, feedback, and education. Monitoring is essential, both in terms of utility bills and in periodic inspection of facilities and equipment. Monitoring provides the basis for feedback to management and staff, enabling them to attend to problems and to maintain the new system and improved operations. CCC provides quarterly written reports to building management; the reports are straightforward and easy to understand, usually just two pages, with a graph highlighting changes in consumption. Staff education is also essential. Operations and maintenance training is necessary to enable building staff to ensure an efficient operation. Staff must also be afforded the opportunity and time to observe and reflect on building operations, to break out of a reactive mode into one in which they are in control of the situation in the building.

Making sure that residents understand what is being installed and how it works is also crucial. CCC generally carries out one or two resident education sessions during the retrofit and construction phase of a project and then provides at least another session annually over the term of a performance contract. Most residents have likely adapted to the pre-retrofit situation, even if it is a very poor one in terms of both comfort and energy use. For example, buildings lacking effective thermostatic control are often overheated, and residents will need to re-adapt after retrofit. In one overheated apartment building, a resident reported, "You can regulate your own heat, as far as I'm concerned. If it's too hot, just open the windows. If it's too cold, shut them and put up the storms" (quoted in Kempton and DeCicco 1985). Different tenants have different adaptations—some of them can be dangerous, such as the use of gas ovens as supplemental space heaters.

Installing temperature-limiting setback thermostats (see Thermostatic Controls, below) may be essential for bringing a building into control. CCC generally uses limits of 75°F for family units and 78°F for seniors units, with a 5°F setback. Such controls can provoke a severe reaction from residents, who perceive it as a loss of control over their comfort and living space. Here is where education comes in, for both residents and management. Residents need to be told what to expect and be given explanations about how the new system operates. Educating management about resident responses is just as important. Kamalay notes some less-than-ideal managers' responses to resident complaints, such as: "If you don't like it, you can leave"; or dishonesty, with managers denying that thermostats are limiting temperatures; or management overriding the limits, defeating the improved controls. Building management and staff must therefore be trained to face the ramifications of such a change in a firm but constructive fashion.

Of course, appropriate attention to technical details is also part of a holistic approach. Such carefulness does not mean getting bogged down in details. Especially since it is crucial to get management onboard as being responsible for achieving the savings, a piecemeal approach can result in details getting "lost in the noise." The auditor should understand all the details and work out a plan to ensure that they are taken care of. CCC's John Snell cites the key rule of thumb that "you can't save the same energy twice." It is important to cross-check engineering heat-loss calculations against billing data and other available measurements, complemented by careful assumptions about indoor temperatures, the seasonal efficiency of mechanical systems (heating, cooling, hot-water), and infiltration and ventilation loads. In very inefficient buildings, it is appropriate to pursue big-ticket items, such as major shell rehabilitation and mechanical system replacements or conversions. These items can help management understand why and how fundamental changes can be made rather than going through a checklist of little items. The retrofit plan should focus on overcoming key barriers, such as energy use information barriers (feedback between paying the bills and operating the system) and up-front costs. Overcoming such barriers can take a lot of resources, and so it is best to try to maximize the energy savings. Ultimately, cash-flow analysis is the determining factor in what gets done. Experience is important; Snell recommends that nascent programs be cautious at first, since confidence in retrofit cost/benefit analysis for a given project ultimately depends on achieved savings in other similar buildings.

System Management, Control, and Behavioral Issues

It is through the management and control of a building's energy-using components that the physical and behavioral aspects of energy use are linked. Approaches for controlling apartment building energy use include unit-level metering, heating cost allocation, thermostatic controls, and energy management systems (EMS). Many questions arise when considering the subject of control, ranging from who pays for energy, to staff training and responsibilities, to specifications for control devices (radiator valves, air conditioner knobs, and thermostats).

It is useful to think about building energy management and control issues in terms of providing feedback. Feedback entails observing or sensing the outcome of a process and feeding the information back to whoever or whatever controls the process. Feedback is crucial at many levels. A thermostat is the archetypical feedback mechanism; it senses the temperature inside a building and provides a signal back to, say, a boiler about whether or not heat is needed. Conservation practitioners know that, if operation is not understood, thermostats can be one of the biggest problem areas in any building. Another level of feedback occurs from residents to the staff, owners, or management of a building. This human feedback loop may be dysfunctional, such as when tenant complaints result in wasteful operational practices by staff seeking to avoid undesirable "feedback." Particularly critical is the feedback (or absence thereof) from those who pay the energy bills to those who make the decisions about building equipment choice and operation. As noted above, the very act of establishing feedback where there has been none can have a profound effect on building operations. Thus, the nature of system management and control depends on the mechanisms (physical and operational) for measuring energy use and comfort and then conveying this information back to a building's owners, staff, and residents.

Unit-Level Metering and Heating Cost Allocation

In buildings where utility usage is measured only by a master (central, whole-building) meter, energy costs are bundled as part of rent. This lack of feedback often causes energy waste. However, attempting to bill tenants directly for energy use raises equity concerns, particularly in low-income settings. The technical possibilities for making residents responsible for energy costs vary with the type of building and heating system. Options include heating cost allocation, submetering, conversion from master- to unit-level metering, and complete system conversion from central to unit-level HVAC systems. To the extent that any of these approaches increase accountability for energy costs, they generally yield energy savings (Judd 1993). Once relieved of direct energy costs, however, building management may shirk conservation measures, such as shell measures, and neglect maintenance items that tenants cannot control but that affect their energy bills. If a conversion involves fuel switching, such as from relatively low-cost gas or oil to high-cost electricity, the result can be unfair to tenants even if site energy use is reduced. Tenant control of energy use in their apartments can also be imperfect because of interactions with neighboring apartments.

Heating cost allocation refers to monitoring or estimating energy consumption in an individual apartment so that the tenants can be billed in proportion to the energy they use. Thorough discussions of heating cost allocation are given by Hewett et al. (1986) and Hewett (1988a), who address five types of monitoring systems. Time meters record the amount of time during which apartment thermostats call for heat. Btu meters measure hot-water flow and temperatures, providing a calculation of the amount of thermal energy delivered to an apartment. Time-plus-temperature systems determine a similar estimate without measuring hot-water flow, which is assumed to be constant when the pump is running. Comfort monitoring systems measure apartment temperature (either actual temperature or thermostat setpoint), to bill for "comfort" provided rather than energy used. Evaporative monitors provide a way to estimate radiator heat delivery. Heating cost allocation is common in Europe, where a variety of systems are found and heating cost allocation equipment is available from a number of vendors.

Hewett (1988a) identifies several issues to be addressed regarding cost allocation, particularly regarding fairness to tenants. First of all, the building should be reasonably energy-efficient; that is to say, heating cost allocation should not be implemented unless cost-effective improvements have been made to the building components and mechanical systems for which the owner is responsible. The metering system should equitably allocate nonmetered costs, assure reasonable levels of accuracy, and be resistant to tampering. Tenant education and disclosure are critical so that tenants understand how the system works, how their bills are computed, how they can identify problems with the system, and what they can do to hold down their energy costs. Budget billing, which spreads costs into roughly even payments throughout the year, may be financially desirable, particularly for low-income tenants and in climates having a strong peak heating or cooling season. However, spreading payments out weakens the feedback, compared with direct monthly energy bills that are responsive to residents' behavior. European countries have developed detailed standards addressing the terms, conditions, and equipment for heating cost allocation; a lack of standards for heating cost allocation is a problem in the United States. The Center for Energy and Environment (CEE) (see Appendix B) in Minneapolis has researched heating cost allocation standards and may be contacted for further information, and a guideline document on the subject has been issued by the American Society of Heating, Refrigerating and Air-Conditioning Engineers (ASHRAE 1994).

Conversions from master- to unit-level metering have been undertaken since the 1970s. Evaluations of these efforts generally showed that energy savings were achieved, although not always through improved systems efficiency or in an equitable manner (Rosenberg 1984; Hackett 1984). McClelland (1983) reported median savings of 14% of total annual gas use in 50 buildings converted from master to tenant metering in the San Diego and Denver areas. Rosenberg (1984) reported a wide range of savings, 3–37%, for a set of conversions in Massachusetts; the lower savings were in town house–style units, and the median savings for the five projects surveyed were 17%. Palermini (1989) examined both heating cost allocation conversions and conversion to unit-level heating systems. Allocation systems showed an average of 16% savings, and system conversions showed much larger savings, up to 73% for a conversion from central gas to individual gas units, although a good part of this very large savings probably came from eliminating the inefficiencies of the old central system. The heating cost allocation experience reported by Hewett (1988a) in Minnesota showed average energy savings of 16% in the first year after conversion and an additional 6% savings in the second year. Electronically based systems, which avoid the need for rewiring while measuring apartment-level electricity use in master-metered buildings, have been demonstrated in New York (Judd 1993). Results show 15–30% reductions in electricity consumption and even larger reductions in demand (residents started turning air conditioners off instead of leaving them on all day during the summer), prompting utility interest in such submetering as part of DSM programs.

The Michigan Public Service Commission has developed policy guidance regarding submetering and conversion from central to unit-level heating and hot-water systems. Conversion from master- to unit-level metering is not allowed under Weatherization Assistance Program (WAP) rules and is also disallowed in some state programs. It is worth reviewing this policy to perhaps accommodate cases in which fuel switching makes sense. There would have to be procedures to ensure that the equity issues are adequately addressed. In particular, landlords should make cost-effective conservation improvements in their areas of responsibility; there should be assistance for tenants in acquiring more efficient appliances and lighting; and landlord agreements must guarantee rent reductions that adequately compensate tenants for increased costs they might bear from the conversion. Although such terms might discourage landlord participation, carefully designed submetering or heat cost allocation conversions should result in reduced costs to both landlords and tenants. Larger savings have been documented with full-system conversions, which have been successfully carried out as noted elsewhere in this chapter.

Thermostatic Controls

Broadly speaking, a thermostat is a device to provide automatic ("feedback") control to a heating or cooling system to regulate temperature. In this sense, thermostats include devices such as thermostatic radiator valves as well as the common wall-mounted thermostat. Energy management systems may have temperature sensors in apartments that control a heating system without any tenant-adjustable mechanisms in the apartment. The ideal thermostatic control system would regulate an apartment's heating or cooling system to provide just the right amount of space conditioning to satisfy the residents' needs. This is not an unambiguous requirement if there are multiple occupants in a living unit, and it may not be a predictable requirement given human whims and their dependence on other conditions, such as the weather and an individual's health. Many thermostats can be set by occupants, letting them control their own comfort levels. Others are not adjustable or adjustable only within limits ("limiting thermostats"), which can be a desirable characteristic from the owner's viewpoint in some buildings. An ideal thermostat would be easily understood and operated by residents. Automatic setback is a desirable feature; significant energy savings can result if heating temperature settings are lowered at night or during unoccupied periods and if cooling settings are similarly raised.

In apartment buildings, nothing regarding thermostats can be taken for granted. In some buildings, thermostats may not exist; in others, they may be nonfunctional or misused. Satisfactory thermostatic control requires the right equipment, properly installed and maintained, as well as residents who understand how thermostats should be used. Kempton (1986) pointed out that many people tend to view thermostats as a valve, to be turned up to add more heat, for example, rather than as an automatic control device. This "folk model" of thermostat operation may be quite functional for occupants. Diamond (1986) found that residents (as well as the building's management and architect!) lacked a good understanding of the thermostats in a seniors apartment complex with heat pumps; thermostat settings correlated poorly with energy use, and some residents preferred to use the manual override that engaged electric-resistance heat.

Approaches for improving the control of apartment building space conditioning depend on the form of the system. Central-heating systems may have central controls, zone control, individual apartment controls, or a combination of these. The controls may or may not involve thermostatic feedback. Unit-level (individual-apartment) systems, including most air conditioning systems, generally have thermostatic controls; such systems are technically simpler to control but still require attention to functionality of the controls and their understandability by occupants and staff.

Control of Central Systems

Some centrally heated apartment buildings were built without any thermostatic control whatsoever. Outdoor reset control systems, for example, sense outdoor temperature and regulate heat delivery accordingly but are not a form of thermostatic control. Examples are controls that link boiler run time, water temperature, or steam valve openings to outdoor temperature. Such systems are inherently difficult to operate in an energy-efficient manner (DeCicco 1988a, 1988b). (Note that outdoor reset control of heating demand is distinct from outdoor reset control of heating system supply temperature in buildings with thermostatic control; as noted elsewhere in this chapter, the latter can be an effective adjunct control strategy in hydronically heated buildings.) Incorporating thermostatic control in buildings that lack it is an important apartment building retrofit strategy, which, of itself, can result in substantial energy savings. Lacking either appropriate thermostatic controls or sufficient understanding of the controls, building occupants develop their own control strategies, which lead to energy-wasteful situations and can defeat attempts to achieve persistent energy savings (Kempton 1986; DeCicco and Kempton 1987; Kempton et al. 1989).

Thermostat placement can be problematic in centrally heated buildings, particularly if resources or the system configuration permit only a few of the apartments to have thermostats. There may be no truly typical location, given the variabilities in exposures, the distribution system, and tenant preferences (and tenant turnover). Whenever the control mechanism responds to worst-case conditions—whether a cold apartment or a complaining tenant—the result is likely to be an overheated building. Auditors often speak of the "double-hung thermostat": an open window that is the only means of temperature control in an overheated apartment. Thermostat placement is particularly challenging in a centrally steam-heated building because of differences in steam travel time to apartments on different floors and in different parts of the building. Thermostats may have internal ("anticipator") settings that are pre-set for single-family houses and so must be adjusted to work properly in apartment buildings. Such adjustments and other control changes may be needed to achieve even heating (Peterson 1984; Katrakis et al. 1986). A strategy for which documented success has not been reported is the use of thermostatic radiator valves.

Installation of multipoint averaging thermostats has met with some success in several apartment building conservation programs (Katrakis 1990). Design questions include the number of sensors needed and where they should be located. One sensor in a north-facing apartment and one in a south-facing apartment on both the top and the bottom floors of a large building may be a reasonable choice. In this case, the thermostats should be pre-set at a desired temperature and protected from tampering so as to prevent individual tenants from improperly controlling the whole system. Another option is a multipoint "cold-spot" locator configuration. A set of thermostats is placed in apartments or other locations prone to underheating. If any one location drops below the setpoint, the boiler fires. This type of arrangement has been successfully used in Minneapolis programs (Lobenstein 1994).

For any type of control system, building management must have a clear understanding of how the system works and what maintenance procedures are needed to ensure effective operation. In a complex building, a thermostat alone is unlikely to provide maximum control and savings. A well-balanced distribution system is also required. Common areas need not be heated to the same temperature as apartment units; such areas may be buffered from building shell losses and less responsive to outdoor weather conditions.

Setback Thermostats

Lowering the temperature setting of a thermostat during the night or when an apartment is not occupied can reduce heating energy use; similarly, raising the setting when full cooling is not needed can cut air conditioning energy use. Thermostats that incorporate a clock to automatically change settings are called "setback" thermostats. Particularly in apartment buildings, automatic setback is more effective than relying on occupants manually setting the temperature back. Nevertheless, effectiveness still requires resident education, with follow-up as well as judicious equipment choice. Use of setback is of dubious value for saving space heating electricity usage in buildings having unit-level heat pumps.

Wilson (1991) provides an up-to-date survey of setback thermostats appropriate for individual dwelling units, discussing the shortcomings of various designs and identifying characteristics of improved models now available on the market. Thermostat manufacturers have been developing improved designs that are easier for people to understand, particularly in terms of setback functions. Apartment buildings with individual (unit-level) HVAC systems can use thermostats appropriate for the system and similar to those for single family homes. A number of thermostat characteristics merit consideration in certain apartment building settings. For example, the elderly have been found to have greater-than-average difficulty with digital systems, particularly those with more functions than a single setting. Elderly residents may find liquid crystal diode (LCD) displays difficult to read and buttons difficult to push, and may be confused by on/off periods set with removable pins (on analog clock thermostats).

Various types of limiting thermostats can be valuable in apartment buildings. These give tenants some degree of control but prevent extremely high settings. CCC has used limiting thermostats with maximum settings of 75°F in family buildings and 78°F in seniors buildings. Also available are "automatic vacancy setback" thermostats, which revert to a lower setting after a given period of time (30 minutes to 12 hours), allowing a resident to obtain higher temperatures when desired but limiting the duration of heating at the higher temperature.

A building's thermal mass affects the way it responds to outside temperature changes and system operation. Thermal mass refers to the capacity of building materials to absorb and store heat. For any building with high mass components, such as masonry, the response to temperature changes is gradual. The resulting time lags must be considered when establishing thermostat setback regimes. High-mass buildings need lead time to bring interior temperatures back up to comfort levels in the morning. On the other hand, thermostats in high-mass buildings can be set back much earlier than they would be in a low-mass structure. That is to say, the boiler can be shut down at 10:00 p.m., and if the building is fairly tight and has a lot of mass, the occupants might not notice a drop in temperature until after midnight.

Education is crucial when installing new thermostats, even those of the best design. From experience with a weatherization project that installed setback thermostats, a practitioner points out: "You need to have two things: a client commitment that they want setback/preheat capabilities plus education about operation at time of installation. You can't install [setback thermostats] and then come around a few weeks later and try to show people how to use them. They will have already soured on the things and they won't like them" (Gill-Polley 1990). In this project, sponsored by the Niagara Mohawk Power Corporation (NY), residents received extensive in-house education, which helped them determine their own thermostat schedule and walked them through the programming operation.

Energy Management Systems

A computerized energy management system (EMS) is a high-tech solution to the frustration of buildings operating out of control. EMS is as much a monitoring system as a control system. Temperature measurement devices are installed in many or all apartment units. The system is balanced or controlled as appropriately as possible. A central computer monitors all parts of the building and assures that there is sufficient heat to meet legal or desired requirements. In some cases, the system merely controls boiler firing. In more sophisticated installations, an EMS can actuate valves, dampers, fans, or pumps to separately adjust heat delivery to various building zones. A key benefit of EMS, however, is that it can inform building operating staff and management of how well the system is working. A manager can identify underheating or overheating without waiting for complaints (which may never come in the case of overheating). Thus, even manual tuning of distribution systems is greatly facilitated. EMS can be set up to allow remote monitoring and control. A manager for an entire housing complex can control several buildings from a central office; an energy service company can monitor the performance of many buildings throughout its service area via a phone link.

EMS systems are relatively costly but can be quite cost-effective when properly installed and operated. In apartment buildings, the retrofitter considering EMS must take into account potential challenges with maintenance personnel and residents as well as routine technical problems of sensor failure or computer malfunction, plus protection against vandalism of expensive equipment. Goldman et al. (1988b) summarized savings from EMS projects in four buildings covering over 2,000 dwelling units, finding average savings of 25% of pre-retrofit energy use and a simple payback of 2.5 (±1.7) years. The EMS results compared very favorably with the 12% median savings reported by Goldman et al. for all system control retrofits. However, a large portion of savings may come from fixing broken controls or establishing a basic degree of feedback control where none has existed; these are often basic elements of retrofit packages involving EMS. Moreover, the robustness of such savings depends on how well the new controls are maintained. For example, Gold (1984) reported nearly 50% savings in the first year following installation of a computerized EMS that provided feedback control in a 159-unit garden apartment public housing complex. Temperature sensors installed in 72 of the 159 apartments were sampled to control a central hydronic boiler, previously designed for outdoor-reset control but having a malfunctioning mixing valve. Thus, using the EMS to establish thermostatic control provided substantial savings, but savings dropped off by the third post-retrofit year, reportedly because of poor maintenance (Goldman et al. 1988b).

A project manager who decides to install an EMS must carefully specify thermostat type and location, properly set up and train staff how to operate the controls, and pay close attention to other balance-of-system issues, such as building envelope integrity. The project manager must also ensure that building management pays attention to operations and energy costs and also educates residents about what to expect with the new system. As part of a performance contract retrofit package for a 150-unit electrically heated apartment building, an EMS contributed 13% toward the estimated total 20% electricity savings (Kamalay 1992). After new thermal windows, the EMS was the second most costly part of this retrofit package, at an installed cost of $140,000 ($930/unit) but had an estimated payback time of 4 years. Resident and staff training were key components of this project, and the arrangements included follow-up contacts to ensure proper operation and maintenance over the life of the contract.

EMS is one means of achieving better control in both apartment and commercial buildings, but achieving good system control and operations is a perennial challenge. Larger apartment buildings can share with commercial buildings the sheer number and complexity of components that are part of the control system. Nearly all such components can be manipulated by the human actors in the building, either staff or residents, in ways that profoundly affect energy use and comfort.

Boiler-Based Systems

A boiler forms the heart of the heating system for many mid- and large-sized apartment buildings. Boilers supply steam or hot water to the distribution system that delivers heat throughout a building. A single central boiler may supply domestic hot water as well as space heat, although domestic water heating is often separately supplied. A variety of types of boilers are found in apartment buildings. The *ASHRAE HVAC Systems and Equipment Handbook* (ASHRAE 1992) is a prime reference among the extensive literature on the subject of boiler types, installations, and operations. The efficiency with which a boiler converts fuel energy (either from combustible fuels, such as oil and gas, or from electricity) to delivered heat is one of the crucial determinants of the overall energy efficiency of a building. A boiler's seasonal efficiency is defined as the ratio of total load to fuel consumed over a year. Methods for characterizing boiler energy losses and measuring seasonal efficiency are discussed by Landry et al. (1993) and Katrakis and Zawacki (1993).

The load on the boiler—that is, the amount of energy it must deliver to meet the building's heating needs—is a key factor in its efficiency. Improvements in a building's shell (for example, tightening, insulation) or heat distribution and control system (for example, thermostat settings, balancing) can greatly lower the load on a boiler—perhaps differently under peak conditions than under average seasonal conditions. Shell, distribution, and control retrofits can save energy but also may decrease seasonal efficiency, lowering the cost-effectiveness of these retrofits if attention to the boiler is neglected. Although the benefits of some retrofits, such as burner tune-ups and leak repairs, can be evaluated in isolation, major boiler retrofits must be evaluated in the context of the overall building because of the close coupling of boiler performance to the "balance of the system" in a building, which determines a boiler's load. Table 3-1 summarizes the energy savings potential from various energy conservation measures that have been tested in apartment buildings.

Table 3-1

Achievable Energy Savings from Conservation Measures for Boiler-Based Heating Systems in Apartment Buildings

Measure	Savings	Payback	Considerations
Tune-up	2–10%	<1 yr	May be a one-time opportunity on conversion boilers
Vent dampers	5–7%	3–10 yrs	Should be electronically controlled
Outdoor reset/cutout controls	7–9%	1–2 yrs	Need to educate building owners and staff about use
Single-pipe steam balancing	0–13%	1–2 yrs	May improve comfort without saving energy
Steam-to-hydronic conversion	18–27%	5–30 yrs	Less costly for two-pipe steam, but also improves comfort and reliability
Adding front-end boilers	6–19%	9–30 yrs	Careful engineering assessment and design are essential

Note: The applicability and cost-effectiveness of any of these measures depend on the particular system and situation in a building. The estimates tabulated here represent what has been achieved in applications that have been prescreened, properly diagnosed, and carefully implemented. Savings are given as a fraction of heating-fuel use, and paybacks are simple, based on ratio of installed cost to first-year fuel cost savings.

Tune-Up

Burner maintenance requirements depend on the type of fuel and burner. Oil-fired systems generally need periodic adjustment; some burners may be greatly out of adjustment upon initial inspection. Gas burners are generally pre-set and have less need for adjustment. A variety of maintenance items should be taken care of annually, as discussed in such handbooks as NYCHA (1982) and Yaverbaum (1979), among others. Critical adjustments are needed to ensure maximum steady-state efficiency as well as safe combustion conditions, which can be determined by flue gas analysis. Electronic testers are now available to compute steady-state efficiency by measuring the temperature and oxygen or carbon dioxide content of the flue gases. Older chemical flue gas analysis kits can also provide reliable measurements, given proper technician training and maintenance of the instruments.

Boiler tune-ups typically cost \$80–\$300 (Lobenstein 1989; Katrakis et al. 1992; UCETF/PEO 1992), depending on the type of boiler and the service needed. Although routine cleaning and "tune-up" do not necessarily bring energy savings, they are part of good maintenance practice, especially for oil-fired systems. A pilot project that provided operation and maintenance training to operators—including instruction in cleaning, control adjustments, and tune-ups on-site in their own buildings—showed an average energy savings of 10% (UCETF/ PEO 1992). This report also details the training approach, operation and maintenance procedures, and repairs made in the pilot sample, which covered ten gas- and oil-heated apartment buildings in Portland, Oregon, many of which had poor operation and maintenance procedures prior to the project.

For boiler tune-ups to achieve energy savings, Lobenstein (1989) points out that selecting the right candidate boilers for a "high-efficiency" tune-up is as important as having a properly trained and equipped technician. Conversion boilers (older coal boilers converted to gas) were found to be good candidates for tune-up in the Minneapolis studies. High oxygen (O_2) readings and high stack temperatures can be indicative of potential tune-up savings through reduction of excess air. On the other hand, tune-ups of boilers designed for gas have shown marginal savings at best (Ewing et al. 1988; Lobenstein 1989). In all cases, testing and adjustment must be conducted under actual operating conditions (for example, if a boiler room door is normally closed, it must not be left open during testing and adjustment, as this can affect the draft on atmospheric systems). With properly trained technicians and screening of candidate boilers, tune-ups yield cost-effective savings. Savings of 6% with six-month paybacks have

A typical Minneapolis area boiler that had been converted from coal to natural gas, fired with an atmospheric burner.

been demonstrated for conversion boilers (Lobenstein 1989), but this
is generally a one-time opportunity. Other programs have shown that
improvements of 2–4% are achievable (Peterson 1983; Lobenstein et al.
1986; Lobenstein 1989). The UCETF/PEO (1992) pilot project, involving tune-ups, operator training, and operational improvements, found
savings of 10% at a cost of $850 per building (training costs included).

Vent Dampers

Vent dampers provide a way to seal the flue when a furnace or
boiler is not firing, thereby reducing the air flows through the boiler
that cause "off-cycle" flue losses. Because losses are highest just after
the end of firing, when the system is hottest but when venting of the

71

combustion products is no longer needed, the controls closing the damper must be precisely timed; the damper must also provide a tight seal. Adequate control accuracy is practically impossible with thermally activated (bimetallic) dampers. The effectiveness of vent dampers also depends on the significance of off-cycle flue losses as part of the overall contribution to system inefficiency. For example, properly sized burners and power burners with otherwise well-sealed combustion chambers or systems with insulated chimneys and other factors tending to reduce boiler stack effects may benefit little from vent damper retrofits. A good review of the use of vent dampers in apartment buildings is given by Hewett (1990), who notes that the ability to recommend vent dampers is hampered by lack of information on operating losses and seasonal efficiencies of apartment building boilers. Vent dampers are more likely to be an effective retrofit for massive older boilers, but even in these cases, test results are variable.

A number of early audit procedures recommended vent damper retrofits largely on the basis of engineering analyses predicting reduced flue losses; however, the considerations noted above are difficult to model and were often neglected in developing such recommendations. In fact, the experience with vent dampers has been mixed. For electronically controlled dampers, reports indicate average savings of 7% of heating energy use with a six-year average, but highly variable, payback (Goldman et al. 1988b). In Minneapolis, field trials of vent dampers, for both water and space heating boilers, have been conducted in a number of buildings and in combination with a variety of other retrofits. The marginal cost of electronically controlled integral flue dampers was $500–$600 of a total installed cost of $2,600–$2,800 for the 199 kBtu/hr–rated water heaters tested by Lobenstein, Bohac, Staller, et al. (1992). Even for systems in which savings (expected to be about 5%) are likely, paybacks may not be quick enough for apartment building owners to consider the retrofit worthwhile. However, in systems in which a water heater and boiler share a common flue, installing a damper on the water heater vent may be needed to obtain good performance of the boiler damper, and measured savings in these cases are usually significant (Hewett 1990).

Thus, current guidance is that thermally activated vent damper retrofits are probably not worthwhile. The value of electronically controlled dampers can only be assessed as part of a careful diagnosis of a given boiler system. Other measures, such as combustion chamber sealing or installing high-efficiency burners, may reduce the incremental benefits and therefore the cost-effectiveness of vent damper retrofits.

Mary Sue Lobenstein

Boiler room showing insulated main boiler with front-end modular boiler installed by EERC, St. Paul.

Boiler Room Combustion Air Dampers

Particularly in older buildings, central gas- or oil-fired units are located in an unconditioned space in the basement with a window for combustion air intake. Although this intake is necessary when the units are firing, it is not needed during off-cycle periods and can result in higher standby losses due to cooling of the ambient air and increased chimney stack flow. Installation of a damper on the window, interlocked with the burners, can allow the damper vanes to be open when a burner is firing and to shut a short while after the burner cycle is completed. Such combustion air dampers may be appropriate for large central boiler equipment.

Boiler Replacement

In many instances, apartment buildings still have their original, decades-old boilers. For example, at least 40% of the boilers in Chicago apartment buildings are brick-set models 70 or more years old, converted from coal to oil and then to gas (Katrakis et al. 1992). Due to the capital requirements of boiler replacement, the decision to replace a boiler must be made in the context of comprehensive retrofit planning, accounting for other aspects of the building as a system and the financial resources available, as discussed earlier in this chapter. In addition, a number of technical questions related to boiler performance should also be answered as part of a comprehensive audit of apartment buildings whose energy use by water or steam boilers is a significant contributor to operating costs.

A key parameter is seasonal efficiency: that of the existing boiler as currently operated and as expected with any feasible tune-up, repair, or modifications, as well as that expected from a replacement boiler. Unfortunately, measurement-based estimation of seasonal efficiency is not yet a routine task. Lack of standard test procedures makes it difficult to compare results and therefore difficult to estimate the energy use impacts of boiler replacement. From the research to date and experience with past retrofits, a number of programs have developed guidelines regarding what levels of efficiency can be expected from common boiler configurations in a given region. Generally, the trend has been toward more aggressive use of boiler replacement as part of a major retrofit package. Katrakis et al. (1992) found that replacement of older equipment with high-efficiency power-burner units is generally cost-effective for all but the smallest (less than ten-unit) apartment buildings.

Many of the fuel-heated building retrofits reported by CCC (1994) involved replacement with fully controlled high-efficiency condensing gas boilers utilizing hydronic distribution. CCC started installing condensing boilers in 1986 and also uses them in electric-to-gas (hydronic) heating conversions; outdoor reset controls and limiting thermostats are standard in all installations. According to John Snell of CCC, condensing boilers may have a long payback time, but experience has shown low seasonal gas use over the life of the equipment. Thus, the favorable post-retrofit cash flow may justify use of condensing equipment as part of a performance contracting package, given a long-term view. According to Snell, condensing operation has been observed under many weather conditions, and there are maintenance benefits from controlled combustion in equipment designed to take condensing loads. However, further research is needed to estimate in-use seasonal efficiencies, to analyze energy savings by source (boiler versus controls, and so forth), and to develop design guidelines for maximizing the benefit of condensing equipment (relating to radiator sizing, pump sizes, and hydronic system temperature setpoints).

Front-end boilers can provide a degree of modularity to hydronic space-heating systems. Installed in parallel with a large existing boiler sized for peak loads, a high-efficiency unit can be sized for average mild-weather loads. The original larger boiler then operates only when the front-end boiler cannot meet the load, resulting in a higher

Front-end boilers installation at the 727 Front Street Building, St. Paul, Minnesota (see case study later in the chapter).

seasonal efficiency. In Minnesota, for example, Robinson et al. (1988) found that most space heating energy use occurs in moderate weather (above 30°F), so that a front-end boiler sized for 25–50% of peak heating load will meet 60–90% of the annual heating load.

Robinson et al. (1986) reported average measured gas heating savings of 15% for front-end boiler retrofits in two 12-unit hydronically heated apartment buildings. The retrofits had simple paybacks of 10–12 years, with better cost-effectiveness for the building in which the front-end boiler was also used for domestic hot water. In subsequent work, Lobenstein et al. (1990) examined savings for front-end boiler retrofits in eight additional apartment buildings. Results were mixed: five of the eight buildings showed savings of 6–19%; two buildings showed no savings; and one building showed higher gas use in the first post-retrofit year and savings in the second year. A number of factors were identified in relation to pre- and post-retrofit conditions that affected front-end boiler retrofit performance.

Pre-retrofit factors that lowered savings included the presence of power (rather than atmospheric) burners, the degree of oversizing, and modular operation of multiple existing boilers. Thus, although use of front-end boilers can be beneficial, careful financial analysis and design of the retrofit, based on a good diagnosis of existing system efficiencies and operating conditions, is needed. Additionally, Lobenstein et al. stated that achieved savings also depend on proper maintenance and operation, including training of operators, for systems retrofit with front-end boilers.

Distribution Systems

The past decade has seen much progress in improving the control and design of boiler-based (steam and hydronic) heating distribution systems, largely through research done in conjunction with apartment building retrofit work by the Minneapolis Energy Office (MEO) (now the Center for Energy and Environment [CEE]) and the Center for Neighborhood Technology (Chicago). Improving the balance of space heating distribution systems helps ensure adequate heat in apartments, improving tenant comfort and allowing lower average indoor temperatures to be maintained, and can improve system efficiency, resulting in energy savings. Approaches to balancing are dictated by the type of system (hydronic or single-/two-pipe steam), type of boiler, layout of the building, and the existence or feasibility of apartment-level controls. Converting two-pipe steam systems to hydronic operation can be a cost-effective retrofit, as illustrated by the case study below.

Single-pipe steam systems are common in the Midwest, Northeast, and in many other areas throughout the northern United States. Single-pipe steam is generally restricted to smaller apartment buildings and is particularly common in "walkup" apartment buildings. Although it may be found in buildings ranging up to 50 units or so, the low-pressure operation inhibits its use in large, tall buildings. Some single-pipe steam buildings can be converted to hydronic heat, and such a major retrofit can be attractive if financing is available. However, the $3,000–$5,000 per-unit cost and ten-or-more-year payback time for conversion of single-pipe steam often exceeds program resources and guidelines. More limited retrofit approaches that improve the balance of a single-pipe steam system and cost less than $1,000 per unit can achieve modest savings (Peterson 1984, 1985, 1986; Biederman and Katrakis 1989). Such approaches are based on the recognition that uneven heating is frequently the main cause of inefficiency in single-pipe steam buildings.

Better balance of single-pipe steam systems can be achieved by

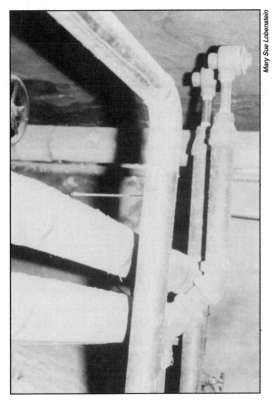

Mary Sue Lobenstein

Oversized air vents are installed at the end of a main line for balancing a single-pipe steam heating system.

diagnosing steam flow in the system and changing vent sizes; this affects delivery to different parts of the building as well as boiler cycling. Katrakis (1989b) identifies four key measures applicable for single-pipe steam retrofit: thermostatic control, larger main-line air vents, radiator vent treatments, and boiler replacement. Balancing work must be done with an understanding of temperature distributions in the building and accounting for stack effects, which can interact with the intermittent heat delivery of single-pipe steam. With temperature measurements as a guideline, temperature variation among apartments can be reduced by use of a proper thermostat, upgrading of main-line air vents, selective replacement of radiator vents, and shell tightening measures. Well-designed single-pipe steam-balancing retrofits have yielded energy savings; for example, savings averaged 10% of total building gas use with a median payback of 1.3 years in a Minneapolis pilot study (Peterson 1986). Sometimes, however, the result is enhanced tenant comfort without energy savings (Katrakis 1989b, 1990).

Hydronic (hot-water) heating systems have a constantly circulating main line with thermostatically controlled zone valves or pumps feeding individual loops for each apartment. This situation approaches the ideal for central heating, since apartments can get just the heat needed, with little delay, and properly calibrated boiler controls can vary the amount of heat delivered to the apartments.

Outdoor reset and cutout controls are relatively low-cost measures applicable to many central hydronic space heating systems in apartment buildings (Hewett 1988c). Outdoor reset controls lower the temperature of the water delivered by the boiler to the distribution system as outdoor temperature rises. Cutout controls turn the boiler off when the outdoor temperature exceeds a certain temperature. By avoiding unnecessary boiler firing in milder weather, such systems result in higher seasonal efficiencies; however, such controls are not applicable to all boiler types. The Minneapolis Energy Office (MEO) tested these controls in medium-sized (12- to 45-unit) hydronically heated apartment buildings in 1982. Controlled experiments demonstrated average savings of 18% of annual gas heating costs (Hewett and Peterson 1984). The reset controls had to be adjusted after installation to address tenant complaints about lack of heat; once adjusted, complaints were minimized, and system operation was improved through reduction of

Mary Sue Lobenstein

A common electronic outdoor temperature reset control with a built-in cutout, as used to improve the efficiency of hydronically heated buildings retrofitted by CEE in Minneapolis.

overheating in apartments and common areas and by increased seasonal boiler efficiency. Subsequent work verified energy savings through a pilot project covering 18 apartment buildings in which reset and cutout controls were installed and operations were explained to building owners but no further follow-up was done. Measured savings ranged from 0.5% to 18% of total gas use, with an average savings of 9% for average installed cost of $615 (1986$) and payback of 1.9 years (Peterson 1986; Hewett 1991). Average 7% savings were also demonstrated in separate tests in Wisconsin (Hewett 1991). Hewett offers guidelines regarding controls specification, installation, and education of building owners/operators that can be followed to assure good energy savings.

CASE STUDY
Converting Two-Pipe Steam to Hydronic Heat

The Center for Energy and Environment (CEE)* in Minneapolis has developed an effective retrofit approach for converting two-pipe steam-heated buildings to hydronic heat. After evaluating a number of conversions, developing detailed retrofit guidelines, and running a successful pilot program, CEE concluded that conversion offers a potential for gas savings of 25% or more with high owner satisfaction and easier post-retrofit maintenance (Lobenstein and Dunsworth 1990; Lobenstein et al. 1993). Converting single-pipe steam systems is much more difficult than converting two-pipe systems because of the need for extensive distribution system replacement. The post-retrofit operational benefits and energy savings are similar, but the substantial added expense for converting single-pipe systems pushed the median payback period to 33 years, compared with 9 years for two-pipe system conversions.

According to Mary Sue Lobenstein of CEE,[†] two-pipe steam with gravity return was a fairly common design for small and medium-sized apartment buildings in a number of northern cities. Larger buildings (for example, over 40 units) may also have two-pipe steam, which has distribution advantages in taller buildings, but these tend to use mechanical return systems (such as vacuum pumps). Candidate buildings were identified during the course of apartment building audits run by CEE. These buildings presented few options for shell retrofit, and so mechanical system retrofits were identified as the major opportunity for energy

---

* CEE was formerly the Center for Energy and Urban Environment and, prior to that, the Minneapolis Energy Office (MEO). Some of the work reported here was done by MEO, but we use the current name, CEE, in all references to avoid confusion.

† Much of the discussion in this case study is based on phone interviews conducted by John DeCicco with Mary Sue Lobenstein of CEE in Minneapolis.

Mary Sue Lobenstein

A six-unit Minneapolis apartment building that was converted from two-pipe steam to hydronic heat.

savings. CEE was able to add wall insulation in one smaller building, but Lobenstein noted that this was a rare opportunity.

The experience of CEE and others reveals a number of difficulties with conventional approaches to retrofitting two-pipe steam, short of system conversion. Mary Sue Lobenstein notes that a common problem in such buildings—as with many centrally heated apartments—is lack of system balance, resulting in some apartments being overheated while others are underheated. There is generally only one central thermostat, often in a ground-floor apartment or sometimes in the boiler room. Balancing such a system requires properly functioning steam traps. A conventional approach might dictate replacing all of the steam traps; however, this is an expensive procedure, and even if better balance is obtained, benefits are likely to be short-lived because of lack of maintenance. Distribution losses also tend to be high in such two-pipe steam systems, particularly when the heating load is low and the high temperature of the steam is largely wasted. Moreover, the higher operating temperature generally results in steam boilers having a lower seasonal efficiency than comparable hot-water boilers.

In Minneapolis, most smaller apartment buildings were found to have poorly trained boiler operators, often people who know just enough to get a license—and qualify for a rent break in one of the owner's buildings. Good maintenance might be seen in larger steam-heated buildings, which typically have a trade-school-trained operator and good management. CEE's experience indicates that establishing better maintenance is wishful thinking in smaller, lower-income buildings. Referring to this stock, Lobenstein notes, "All two-pipe steam buildings have poor maintenance." Expense and high maintenance requirements also make thermostatic radiator valves a poor choice for retrofit.

Earlier experience identified many of the potential benefits of converting two-pipe steam to hot water (Lobenstein et al. 1986). CEE then developed a set of guidelines and specifications for carrying out such conversions (Lobenstein and Peterson 1988). A pilot project was conducted on five two-pipe steam buildings in 1988–1989. System conversion need not entail boiler replacement—in three of five buildings evaluated for the pilot, existing boilers were converted from steam to hot water. This was the case in the "2200" building, a six-unit, 8,000-square-foot walkup structure originally built in 1913 but which had a recently replaced cast-iron boiler only four years old.

The CEE auditor examined the "2200" building and determined that it was a good target for conversion. CEE obtained the owner's interest and, with the program providing technical assistance and a 10% rebate toward the retrofit cost, obtained the owner's consent for a conversion project. CEE wrote the specifications for the job, provided a list of suitable contractors, and assisted in bid selection. This type of technical assistance is of prime importance for owners—it helps overcome a major barrier, namely, that owners often lack technical information and confidence that the right work will be correctly done. The boiler was converted from steam to hot water, and the distribution system was converted by removing traps and upgrading valves. New controls were installed, consisting of remote sensing thermostats and an outdoor cutout control. The thermostat sensors were placed in two apartments and operate the boiler whenever one of them goes below the setpoint. Total retrofit cost was $6,800 in March 1989, when the work was performed.

CEE's conversion guidelines call for either repacking or replacing steam valves throughout the system. Leaks that are not apparent when running with low-pressure steam become obvious with the higher pressures of a hydronic system. Given the generally poor maintenance in apartment buildings, valves are rarely in good shape. In building "2200," the owner wanted his own maintenance worker to take care of the valves, hoping to save contracting costs. But this setup created delays, and the work was done incompletely and incorrectly, so a contractor ultimately had to redo all the valve work. CEE's Lobenstein points out a lesson here, namely, that it is risky to count on what owners might say they will handle themselves for a critical part of major retrofit projects, such as a system conversion.

In many steam-heated buildings, apartments in the basement or near the boiler room have few radiators since hot steam pipes running through these locations can provide adequate heat. After conversion to hot water, which operates at a lower temperature, such apartments may lack sufficient radiation to meet their load. CEE recommended adding radiators to the basement apartments in building "2200," but the owner demurred. Sure enough, tenants in the basement apartments complained about insufficient heat. The owner then installed the needed radiators, which fixed the problem. But the result of this delay

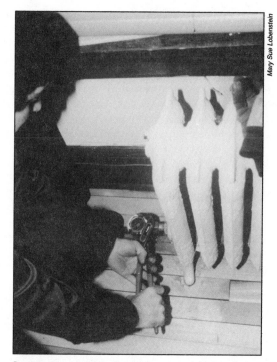

Contractor opening the steam trap of an existing radiator to remove the trap mechanism, for conversion to hydronic use.

was an initially mixed tenant satisfaction response to the conversion. In another building of the pilot program, the need for additional radiation was not anticipated but turned up after retrofit when basement tenants said their apartments were too cold.

Evaluations of CEE's five-building pilot steam-to-hydronic conversion project showed average savings of 25% of pre-retrofit weather-normalized total building energy use. The savings were largely the same whether or not the boiler was replaced or converted, indicating that the benefit is from system operation rather than a better boiler. Benefits are due to more even heating, reduced distribution losses, and reduced boiler losses (since the boiler operates at lower temperatures when heating water instead of making steam). The simple payback for the "2200" building retrofit was 11 years, longer than the pilot program median payback of 6 years. This retrofit is still cost-effective for a gas-heated building, considering the long life of the system, with an estimated cost of conserved energy of $3.40/MBtu.

Factors critical to the success of the program are (1) technical assistance, including detailed bid specifications, and (2) careful pre- and post-retrofit surveys of the buildings and tenants, including follow-up to ensure proper operation and to address problems (for example, the need for additional radiation). Also, a 10% rebate of conversion costs induced participation by several building owners who would not otherwise have undertaken the retrofit. The written bid specifications and other technical services provided by CEE were indispensable to the building owners and were also valued by the contractors who performed the conversion work. A full discussion of the program, including attached survey forms and a sample bid specification, is given by Lobenstein and Dunsworth (1989).

Electrically Heated Buildings

As noted in Chapter 2, 43% of apartment households use electricity as their primary source for space heating. A major division is between systems with and without air ducts. Ducted systems include various forced-warm-air designs, which may include air conditioning, as well as heat pumps, which provide both cooling and heating (generally with electric resistance as the second stage). Although ducted systems account for 57% of electrically heated apartment households nationwide, these are, unfortunately, the systems about which the least information is available. Apartment buildings with ducted systems are typically newer and have been less frequently targeted for retrofit. Better information is available about systems using some form of unit-level electric-resistance heat, which account for 42% of electrically heated apartment buildings. Few electrically heated apartment buildings use hydronic distribution.

Because nonducted electric systems avoid point-of-use energy conversion and distribution losses, the main opportunities for energy savings are found in reducing building losses and system loads. Such opportunities may include thermostat setbacks; shell measures such as window retrofit, insulation, and air sealing; pipe and duct wraps; water heater insulation and water conservation devices; and efficient lighting retrofits. It is also beneficial to ensure that apartments are adequately isolated from one another, perhaps through insulation or air sealing of adjoining walls and floors. These measures are treated in separate sections throughout this chapter. A number of conservation programs targeting electrically heated apartment buildings, such as those of Seattle City Light and the Hood River Project, have focused successfully on load reduction retrofits.

Nonducted Resistance Heating Systems

The relatively high cost per supplied Btu makes electric-resistance heat a poor consumer choice except in areas of low electricity cost, such as hydropower regions like the Pacific Northwest and parts of Canada. Electric resistance is often selected because its low installation cost is appealing to developers, landlords, and building managers. In newer buildings, individual apartment metering is easy and commonplace. Such buildings have the advantage of greater tenant responsibility and the disadvantage of reduced owner incentive for efficiency improvements. On the other hand, if tenants pay their own electric bills, they have a cost motive to conserve in the ways that they can, for example, thermostat setback, paying attention to door and window closure, and reporting problems to the building management. Thermostat setback control is often easy to implement, and in many systems, individual rooms can be independently controlled. Giving tenants the ability to control their own comfort is generally advantageous. On the other hand, building management lacks a cost motive for conservation. This situation is problematic in many electrically heated buildings since the greatest opportunities for conservation are often through shell retrofits that must be paid for by the landlord.

In practice, retrofit of electrically heated apartment buildings can be quite challenging. The initial conditions are often poor: low-cost baseboard strips are frequently installed in substandard wall locations with cheap mis- or noncalibrated thermostats, inadequate ventilation, and no insulation. Tenants may be paying more for electricity than they pay for rent (Hayes 1992). For audit and diagnosis, the effort involved in metering individual units presents its difficulties but does provide the opportunity for unit-level diagnosis. In Hayes's (1992) audit of a four-story, 42-unit apartment building in upstate New York, analysis of electricity consumption histories revealed high electricity usage in corner apartments and the top floor, indicating shell losses. A comprehensive physical audit is also needed, including thorough shell diagnostics but also paying attention to the electric heat components. Broken, malfunctioning, miscalibrated, or incorrectly wired heating unit controls are likely to be found.

Electric baseboards are common because of low initial cost, ease of control, and inherent efficiency. Because baseboard units deliver heat primarily through convection, free air flow (unconstrained by furniture or window curtains) is essential for heating efficiency as well as fire safety. On the least expensive systems, each baseboard housing has an uncalibrated thermostatic control knob, which cycles the elements on or off. Slightly more sophisticated is a system comprised of the same delivery unit but with a centralized (usually room-level) thermostat, again usually uncalibrated. Such a system allows isolated temperature control by the tenant and, with frugal habits, can result in significant savings. Slow response times, however, may discourage manual setback. These installations are easily retrofitted with calibrated digital setback thermostats, but these must be installed by a licensed electrician in most areas. When all baseboards in an apartment are controlled by a central thermostat, a setback unit would most likely be cost-effective. However, use of setback thermostats may not succeed if leaky envelope conditions are not also addressed.

Electric radiant space heaters are either portable space heaters or wall-mounted radiant panels. Radiant heaters are designed to operate at higher temperatures than baseboards and deliver much of their heat through radiation rather than through convection. Quartz-type radiant heaters are a subset of this group and operate at very high temperatures. Radiant heaters are likely to be tenant-owned back-up units. Advantages are rapid response time, allowing for deeper temperature setback, and use as "people heaters," providing comfort at a lower overall apartment temperature. On the other hand, these devices can present burn or fire hazards, and reliance on such heaters is not a prudent design strategy for apartments.

Radiant ceiling panels are low-temperature resistance strips either imbedded in the plaster ceiling or hung from the ceiling joists before sheetrock installation. They are usually controlled by line-voltage thermostats located in each room. Slow response times make manual setback regimes unattractive. High insulation levels and attention to convective bypasses in spaces between the floor and ceiling are essential. Workers must be cautioned when cutting or drilling these surfaces, since irreparable damage to resistance coils can occur from careless work. Radiant floor slabs, usually found only in single-story apartments, have resistance coils either imbedded in a concrete floor slab or laid below it. Due to the high mass associated with their design, these systems have exceedingly slow response times, making any sort of setback regime virtually impossible. Though warm floors can be very comfortable, perimeter heat loss can be extensive unless the slab is well insulated; exterior foundation perimeter insulation may be an important retrofit.

Electric convective space heaters can be either free standing or wall mounted, usually on an exterior wall, often under a window. Both deliver fan-forced warm air, providing a fairly quick response time. Usually, temperature controls are integral to the unit; thus, automatic setback retrofits are not possible. Wall-mounted units usually extend completely through the wall, much like air conditioners, and have outside air intake options; through-the-wall heat pump installations are also becoming more common. Such configuration can have high losses to the exterior from conduction through the metal casing and from air leakage, implying opportunities for sealing retrofits. Basic maintenance includes filter cleaning and fan lubrication.

Ducted Electric Heating Systems

Ducted systems are often problematic even under the best of conditions, such as those found in higher-income single-family residences. In a low-income apartment building, ducted systems can be very poor performers. Proctor (1993) warns of many difficulties regarding misinstallation and poor performance of ducted systems in apartment buildings and advises that such designs be avoided when possible. Nevertheless, 3.5 million apartment households have some form of ducted electric heat, either central air with resistance elements or heat pumps. In a sample of recently constructed buildings, including several apartment complexes, Hammarlund et al. (1992) found high incidence of below-specification air flows and refrigerant overcharge in heat pump systems, suggesting potential savings of 18% for cooling and 19% for heating; measured savings results are yet to be reported, however. Compared with detached single-family applications, problems are to some degree lessened when the HVAC equipment and ductwork are completely inside the apartment. In such cases, Modera (1993) points out, a system may have an acceptable level of performance, given adequate routine maintenance (which is, of course, difficult to ensure in some settings). Moreover, duct systems are repairable, and progress has been made in techniques for diagnosis and duct sealing (Cummings et al. 1993; *Home Energy* 1993; Palmiter et al. 1994).

Ducted systems, depending on their layout, can have several problems, including poor insulation, air leakage, and lack of balance (which can cause indoor air quality problems). When furnaces or ducts are located in unconditioned spaces, such as a basement, crawl space, or attic, extreme care in duct sealing and high levels of duct insulation are essential. Even if ductwork is contained entirely within the thermal envelope of the individual apartment, heat loss through ductwork to ambient apartment air may leave registers at extreme ends of the system with insufficient heat. This example points to the importance of proper installation of ducts during construction. Added care in labor is required, perhaps adding another two hours to a typical job (Proctor 1993). However, this additional installation cost is inexpensive compared with the avoidable energy expenses and the cost of repair after the fact, when access may be difficult or impossible.

Electric furnaces combine the high energy costs of resistance heat and high distribution losses of duct runs in what is often a very poor system. Design air temperatures for electric furnaces are not as high as those from a combustion furnace. To compensate, air flow rates from an electric furnace are often greater than in conventional units. Heat loss as the air flows through uninsulated ductwork can make for uncomfortable, drafty conditions and insufficient delivered heat. Also, warm-up time after a setback period is longer, requiring greater anticipation of required heating needs.

Heat pumps are found in 15% of electrically heated apartment households, primarily newer apartments in milder, Sunbelt regions. Air delivery temperature issues are even more serious for heat pumps than for electric-resistance furnaces. Because the temperature increase across a heat pump coil is so low, careful attention to air sealing, duct insulation, and register placement is essential. Ducts located in unconditioned or buffer spaces—a foolish design—may suffer serious conduction as well as leakage losses during both the heating and cooling seasons.

Few measurements are available of retrofits for ducted electric heating in apartment buildings. Side-by-side tests of relocating ducts into conditioned space in two apartments of a four-plex apartment building showed dramatic savings of 34% of heating energy and 71% of cooling energy use (Guyton 1993, discussed by Palmiter et al. 1994). Evidence from studies of single-family residences is applicable if one keeps in mind that duct retrofit requires great skill and care, including "special training in repair methods, advanced leakage diagnostics, and skilled use of blower doors, duct tests, and micromanometers" (Palmiter et al. 1994, 3.186). The exact nature of losses and leakage must be properly diagnosed with an understanding of a system's physics, as well as operating economics, to determine whether retrofit is worthwhile, and if so, exactly what should be done. For example, high measured duct leakage does not justify repair work if most leakage ends up inside the living space.

Routine maintenance for ducted systems includes filter cleaning and replacement; cleaning of coils, elements, blowers, and registers; and ensuring proper functioning of controls. Although the energy savings from these measures are small or nil (field results are not available), these maintenance items are important for tenant comfort, indoor air quality, and safety.

Thermostat setback is of dubious value with heat pumps. Special, staged thermostats are needed to avoid use of the resistance elements during warm-up periods, which are lengthened because of the low heat delivery; comfort can be so compromised that setback usage may be unreliable and cause occupants to frequently override the system.

The replacement of ducted electric-resistance systems with heat pumps is a largely unexplored retrofit option. Air-to-air heat pumps can show significant improvement over resistance heat, even in northern climates, since heat is needed in many mild months, when a heat pump can have a high coefficient of performance. However, the applicability of such a retrofit in the apartment building context may be limited; we are unaware of evaluations reporting experience with this type of retrofit.

Groundwater heat pumps are even more efficient than air-to-air heat pumps and can be cost-competitive with natural gas in some localities. Until recently, however, high installation costs have inhibited their use in apartment buildings. In 1993, the Department of Housing and Urban Development (HUD) installed groundsource heat pumps when rehabilitating the 348-unit Park Chase Apartments in Tulsa, Oklahoma. Technical assistance was provided by the local utility, Public Service Co. of Oklahoma, which also provided financial incentives for the new equipment. The utility recommended for HUD "to do the most you can when making conservation improvements" because of the initial equipment cost savings and the ongoing future energy savings (Henderson 1992, 1). Conversion to unit-level metering and comprehensive shell retrofits enabled downsizing of the preexisting 720-ton chiller system by 181 tons, reducing new equipment costs substantially. Additional benefits cited for the system include reduced maintenance, avoidance of vandalism, and projected energy savings of 45% compared with the old system; post-retrofit analysis was not available at the time of this writing (ClimateMaster 1994). In California, a pilot demonstration of groundwater-source heat pumps as a retrofit for apartment buildings is currently underway in which the Sacramento Municipal Utility District and the Sacramento Public Housing Authority are installing groundwater heat pumps at two apartment building sites. Again, part of the attraction of this system is reduced risk of vandalism, since exterior system components are located underground.

Shell Retrofit

Shell, or building envelope, retrofit can save energy in some types of apartment buildings if its limitations and interactions with heating systems are understood. In some buildings, the synergism of shell retrofits bringing loads under control along with high-efficiency mechanical systems can yield large savings. For a number of building types, however, shell retrofit is inherently difficult and sometimes impossible. For example, masonry walls cannot be easily insulated, and flat roofs often have limited space for insulation. Windows are often left as the only or "obvious" opportunity for shell retrofit, but experience with apartment building window retrofits has often shown poor results. The classic story is that of window retrofits installed in centrally heated buildings without addressing the heating system and resulting in high expense but zero energy savings (Judd 1993). Shell retrofit has a somewhat better track record in electrically heated buildings, where it is often the main option available; however, payback times can be long. Here we review general energy savings evaluation results for apartment building shell retrofit and then discuss approaches for addressing this aspect of efficiency improvement.

Goldman et al. (1988b) found that window retrofits yielded an average measured energy savings of 12% with an 11-year payback in both fuel- and electrically heated buildings. They discovered that other shell retrofits (primarily insulation) in electrically heated buildings yielded an average 14% measured energy savings with a 23-year payback. The Hood River Study (Hirst et al. 1987) examined shell retrofits in 41 electrically heated apartment buildings, mostly low-rise structures less than 20 years old and already partly insulated. Measures included, among others, R-38 basement or crawl-space insulation and triple glazing, installed at an average cost of over $1,500 per unit and resulting in a 14% average energy savings with a 15-year average payback. The Seattle City Light (SCL) program also addressed low-rise apartment buildings with a set of shell, water heating, and lighting retrofits (Okumo 1991; SCL 1991b). Savings ranged from 4% to 18% of pre-retrofit consumption, giving 21- to 35-year payback periods, with window retrofits accounting for roughly 70% of the savings (see the Seattle City Light Retrofit Experience case study, below).

More recently, it has been recognized that shell work can be an attractive part of a major retrofit package for both fuel-heated and electrically heated buildings. Without a reliable building shell, it is virtually impossible to bring mechanical systems into control (Kamalay 1992). When structurally feasible, envelope measures can work together with mechanical system upgrades, better controls, and better operations to yield substantial overall savings. The Goldman et al. (1988b) survey found that comprehensive packages, including both shell and system retrofits, showed average energy savings of 26% with a 6-year payback. The comprehensive packages delivered more than twice the savings and slightly better paybacks compared with mechanical-system-only retrofits and were clearly cost-effective compared with the marginal performance of shell-only retrofits.

Thus, shell measures can have an important role in upgrading energy efficiency in some apartment buildings. Window replacement is often an attractive option when rehabilitating older properties, adding merits (property value, maintainability, aesthetics) beyond energy savings. Moreover, shell improvements are often essential for bringing building loads into control. The resulting lower building loads can permit a smaller and more efficient replacement mechanical system, yielding a cost-effective overall package with substantial savings. For example, a comprehensive retrofit of an all-electric public housing development in Danbury, Connecticut, which involved an investment of $3,800 per unit and included low-emissivity replacement windows, yielded a 24% electricity savings with a projected 7-year payback (Kamalay 1992).

Windows

Building owners often view the addition of storm or replacement windows as a desirable retrofit. Beyond aesthetic value and capital improvement benefits, window manufacturers also claim large energy savings. Energy conservation program managers may look favorably on window treatments because they can be logistically straightforward, with relatively standard materials, simple bidding, and a high materials-to-labor cost ratio (required in earlier U.S. Department of Energy/WAP rules). For larger structures with all-masonry walls and limited attic access, windows may be the primary architectural element accessible for treatment. Window replacements were once a major component of prescriptive apartment building retrofit programs. As noted above, however, retrofits that include windows but ignore mechanical systems frequently fail to save energy. This experience has led some to think that window retrofits are too costly to be useful for apartment building energy conservation. But both these poles of thinking—prescriptive window replacements versus the belief that window replacements are ill advised for apartment buildings—are flawed in their neglect of the building-as-a-system approach to retrofit planning.

High-quality windows reduce conductive and radiant heat losses as well as infiltration and induced drafts. (See Wilson and Morrill [1995] for a primer on window technology.) By providing the benefits of lower direct heat loss compounded by higher radiant temperatures, energy-efficient windows can improve comfort with the same thermostat setting or maintain it with lower thermostat settings, thus contributing to substantial savings as part of a comprehensive retrofit package. Because window replacements are so expensive, however, careful engineering and financial analysis must be done to evaluate their cost-effectiveness. CCC has included window replacements as part of their performance contracts in a number of major apartment building retrofit projects. In one project involving a $227,000 retrofit package for a 60-unit electrically heated apartment complex, window replacement comprised 66% of total project costs and had an estimated ten-year payback. However, the efficient windows permitted installation of temperature-limiting thermostats, thus contributing 77% of the energy savings with a seven-year simple payback for the project as a whole (Kamalay 1992).

Many retrofit programs that regularly include total window changeouts have developed fairly standardized approaches to specification and installation. New York City developed an excellent specifications and installation manual (NYC/HPD, undated), which can serve as a useful model for other programs. The National Fenestration Rating Council produces a directory of over 2,000 products, comparing certified energy performance data from over 60 manufacturers (NFRC 1993). A number of programs negotiate with building owners for contribution to a retrofit, using window replacement as the quid pro quo for investing in other cost-effective energy-saving measures.

Options for dealing selectively with damaged windows include sash replacement (if a good match can be found) or rebuilding existing sashes and casements (which is labor-intensive). Weatherstripping efforts generally yield very little air leakage reduction or are so short-lived as not to warrant the application. Freeing up stuck sashes and casements, cleaning and adjusting interior hardware, and adding sash latches can be more effective and durable than weatherstripping and may be worth the added labor costs; these are particularly desirable strategies for windows in historic structures.

Dixon (1992) discusses a number of considerations in specifying replacement windows for apartment buildings. The future maintenance and repairability of window retrofits are significant. Vinyl or wood sash windows should be considered, particularly in housing where vandalism and theft is an issue. In high-rise buildings, the interior stack effect can drive moisture to upper floors, resulting in heavy condensation on aluminum frame windows lacking thermal breaks; this effect provides another rationale for replacement with energy-efficient windows. In some situations, leaky windows are the only source of adequate fresh air in an apartment. Replacement with extremely airtight windows may lead to excessive moisture buildup or even hazardous air quality problems, such as carbon monoxide buildup from unvented combustion appliances. Some localities require childproof guards or opening-height restrictions on all replacement windows out of concern for safety in tall structures.

Window specifications should also consider local climate needs and solar gains (Wilson and Morrill 1995). Windows are now available with low-emissivity ("low-E") coatings that can be selected for a given climate and orientation. "Northern" low-E windows transmit the most sunlight while blocking heat loss and are ideal in heating-dominated climates. "Southern" low-E windows help shade unwanted solar gains and are appropriate in cooling-dominated climates. Window characteristics can also be specified for the different faces (north, south, east, west) of a building; careful consideration of the available options is important when making the major investment entailed in replacing the windows in an apartment building. Unfortunately, window suppliers may not be well versed in appropriate specifications; for example, some windows promoted in northern Wisconsin as "solar glazing" had high shading coefficients more suitable for warm climates.

Storm Windows and Movable Insulation

Storm windows were a standard measure in the New York City weatherization program from 1975 through the early 1980s. Window opening behavior, lack of durability, and unproved energy savings led to abandonment of this option in favor of replacement thermal windows (Judd 1993). Storm windows are not an attractive apartment building retrofit except in special situations in which treatment is needed and resource limitations prohibit window replacement. Exterior storm windows can increase comfort in common spaces and first-floor apartments (where they also provide a security benefit) as part of a strategy for bringing central-heating systems into balance (Katrakis 1989a). Interior storm sashes may be used effectively if prime windows are inefficient but in reasonable condition and may be useful in buildings with existing casements or other nonstandard windows; however, durability and summer storage may be an issue. Poor-quality primary windows (particularly aluminum sliders) have been retrofitted by adding a new, better-quality aluminum window on the interior, in effect turning the original window into a storm window (Crockett 1990).

The relative benefits of movable insulation systems (curtains, shutters, or shades) have become less attractive given the commercial availability of improved new window technologies. But if window replacement is not in the budget, and if the tenant is the primary client of the retrofit program (such as in U.S. Department of Energy [DOE] weatherization), then movable insulation may have a place. Benefits accrue directly to the tenants, who might be able to take the shutters or curtains with them should they relocate. Heat-loss calculations for either movable insulation or storm windows may underestimate the savings potential of these retrofits when they allow improved tenant comfort through higher mean radiant temperature; this can translate to energy savings if tenants control their own heat.

The need to deal with leaky, single-glazed skylights is a problem that also comes up in apartment buildings (commonly in stairwells). It is difficult to repair or double-glaze these units, so the benefits of day-lighting may be outweighed by the heating energy losses. One useful option is fabricating a cover to seal off and insulate these units during the heating season. Interior transparent covers for the bottom of the light well can insulate the opening and preserve the daylight. Skylights can now be obtained with sealed multiple glazings, low-E coatings, and low conductance (argon); such advanced versions might be worthwhile when replacement is needed for nonenergy reasons.

CASE STUDY
Window Retrofits in a Chicago Complex

Built in 1926, this four-story (three stories plus basement), 72-unit building was originally the Drexel Residence Hotel in the fashionable Hyde Park area of Chicago.* The structure is solid brick, with a stylish facade and 18-inch-thick firewalls between each apartment. The building is well maintained, and the owner is concerned with energy efficiency and other critical issues (such as lead paint). The owner is proud of the fact that the property, with monthly rents ranging from $300 to $425, provides affordable housing in the Hyde Park area. Townhouses across the street sell in the quarter-million-dollar range.

5220 South Drexel was among the first large, privately owned apartment buildings weatherized by the city. Retrofits included windows for 55 apartments replaced with a 50% owner cost-share, plus windows for 5 more apartments completely paid for by the owner. Due to the difficulty of mustering so many tenants to complete eligibility applications, the weatherization was done in two stages. The first 33 units were completed by December 1992, and the remaining 22 units, plus the 5 done by the owner, by May 1993. The owner concurrently installed smoke detectors and new apartment doors throughout, plus new steel fire doors in the stairways. Along with the window replacements, the new doors helped reduce the stack effect air flows, allowing better temperature control and heating system operation. The new windows are high-quality, double-glazed, vinyl-framed units. In some cases, original pairs of 40-inch-wide windows were replaced with three 24-inch units for both cost and safety reasons. Although these replacements changed the exterior appearance of the building, the result did not detract from the aesthetics of this property. No physical retrofits were made to the central-heating system, but a concerted effort was made to ensure careful system operation to take advantage of the reduced load expected from

* Information in this case study is based on an interview conducted by Tom Wilson with a representative of the firm owning the 5220 South Drexel property.

Tom Wilson

Apartment building at 5220 South Drexel Street, Chicago.

the door and window retrofits. The old 2-MBtu/hr gas-fired boiler that supplies the heat and hot water would be eligible for replacement if the owner contributes 50% of the cost.

Since record keeping for the retrofit consisted only of client (individual apartment) files and the job was done in two stages, it is difficult to determine the total weatherization project cost from the city's records. The client files indicate that each window cost about $90 (materials plus labor) and that each apartment received between four and seven windows, implying an estimated total cost of about $29,000. A weather-normalized analysis of the building's utility bills was conducted for the periods June 1991 to October 1992 (pre-retrofit) and February 1993 to April 1994 (post-retrofit). Both heating and baseload energy use were reduced, with the building's reference temperature rising from 63°F pre-retrofit to 66°F post-retrofit. The estimated total energy savings were 12,500 therms, or 21% of the 58,200 therms pre-retrofit normalized annual consumption. Although energy usage is not closely tracked, the owner perceived that the weatherization effort had saved about one-third of the building's heating load. The annual gas bill savings amounted to $6,300, implying a better-than-five-year payback and a savings-to-investment ratio of 2.7 (discounting at 4.7%/yr over a 20-year measure life).

Such impressive savings from a window replacement are larger than what would be predicted by steady-state heat-loss calculations. It is likely, however, that the building was very poorly controlled previous to the retrofit because of high air leakage rates and extreme stratification from air flows through common spaces. Since the retrofits were complemented by careful heating system operations, isolating individual apartments and restricting exfiltration probably allowed more effective heat delivery. The building owner reported that boiler steam cycles were reduced from about 2 hours to 1.5 hours.

The owner saw benefits in addition to energy savings from the replacement windows, for which he had to contribute 50% of the costs, as opposed to storm windows, which would have been provided at no cost to the owner. These benefits included greater air tightness with fewer drafts, reduced maintenance, improved appearance both inside and out, and the elimination of a major source of lead dust. The owner also recognized that much greater savings might have been available through mechanical improvements. The present boiler is "getting thin on the bottom" and demands $1,500–$2,000 per year in maintenance. The distribution system clearly needs balancing, as the front of the building is still cold while the rear tends to overheat (the boiler is at the rear). Since domestic hot water is also supplied by this boiler, there are probably extreme inefficiencies during summer operation. Because there are no central mixing valves and both hot- and cold-water taps in many of the apartment bathrooms, water can be delivered at scalding temperatures.

The owner is very pleased with the work accomplished, stating, "It's a great program, and it certainly made a difference here. We've had two buildings completed so far and are working on two others. All of the others include boiler work as well." Also, because he is very satisfied with the contractor's installation, the owner has hired the same firm to complete the job on common areas and ineligible apartment windows.

Doors and Vestibules

Of themselves, door treatments (replacement, storm, weatherstripping) generally do not present a significant energy savings opportunity in apartment buildings, and door leakage is often less important than other bypass work. However, attention to doors can be warranted for reasons of air quality, security, fire safety, and aesthetics. Additionally, the condition of individual apartment doors, stairwell doors, and other common-area doors can affect air flow patterns in apartment buildings. Careful diagnosis is needed to determine what if any treatment might be needed; doors leading to common spaces must generally meet fire codes. Heavy, steel-clad, and insulated exterior doors offer modest energy savings and added security.

Storm doors are generally inappropriate for large apartment buildings, even when unit layouts would allow their installation. Adding storm doors is rarely cost-effective because the added thermal resistance is applied only to a small area (perhaps 20 square feet) at a fairly high price. Upgrading and sealing the primary door is generally much more effective for controlling air leakage. Nevertheless, Katrakis (1990) has identified several good reasons for installing storm doors in some apartment buildings: (1) draft reduction, (2) summer ventilation (through a screen door), which can reduce air conditioner use, and (3) increased tenant or owner satisfaction. Another justification for storm doors may be preservation of the primary door for aesthetic reasons. For small row house–type structures, some programs undoubtedly still include storm doors. If storm doors are used in apartment buildings, safety glass may be desirable or even required.

Vestibules can have thermal benefits, as well as provide added security and easy access. Vestibules can create a moderating buffer zone that reduces conductive heat loss and can provide an airlock entry that reduces indoor-outdoor air flow when the primary door is opened. The latter benefit is realized, of course, only when the space between the inner and outer doors is large enough that one can be closed before the other is opened. In tall buildings, a vestibule's airlocking ability can be particularly important because of the substantial stack effect. Although these air flow patterns are probably more effectively sealed at the top of stairwells and elevator shafts, reducing the source of cold air at the bottom also contributes to control.

Insulation

Adding insulation is often less important in apartment buildings than it is in single-family structures for several reasons. One is the relatively greater importance, in apartment structures, of addressing mechanical systems, which are frequently oversized, decrepit, and poorly controlled. Another is the more compact nature of apartment buildings, which typically have less exposed surface area per enclosed volume of dwelling space than single-family detached units. Even in row houses, only the two shortest of the four walls of each apartment (except for end apartments, of course) are exposed to exterior temperatures. In multistory buildings, at least in theory, only top-floor units should need ceiling insulation.

In some buildings, such as those with masonry walls or flat roofs with limited ceiling space, insulation options may be limited by the nature of the construction, making reinsulation prohibitively expensive. Bypasses and interior air flows may result in effective exposure of units to outdoor air, but these problems may be better addressed by selective insulation and sealing measures than by overall wall insulation. On the other hand, when extensive rehabilitation ("gut rehab") is justified for reasons that may include energy waste, there are likely to be opportunities to substantially upgrade insulation in an apartment building.

Because of the enormous variety in apartment buildings, the opportunities for and likely effectiveness of adding insulation must be evaluated as part of a careful audit. Different types of structures present different situations, and the retrofit planner must be knowledgeable about materials use and installation techniques as well as about how insulation interacts with heat, air, and moisture flows. These interactions can be quite complex in an apartment building. For example, adding ceiling insulation above top-floor apartments can exacerbate overheating in these units, leading to window opening, which aggravates the stack effect, resulting in colder lower-floor apartments. The potential for savings from insulation must therefore be carefully evaluated using the building-as-a-system approach. Wilson et al. (1990) discuss many of the considerations involved in specifying insulation upgrades for various apartment building types.

CASE STUDY
Energy-Efficient Building Rehabilitation

Major renovation or rehabilitation ("rehab") of an apartment building provides an important opportunity for changes that can improve energy efficiency at low additional cost. Such opportunities can be lost because the context of rehabilitation efforts is generally different from that of conservation efforts; for example, rehab efforts are often initiated by different parties than those working on energy conservation. The main goal of rehab is to maintain viable and affordable housing—a pressing issue in many areas of the country. In recent years, with relatively low and stable energy prices, energy conservation is likely to be an afterthought when designing rehab projects. Rehab can involve replacing nearly everything but the structural components of a building, including new interior surfaces, doors, windows, and mechanical equipment. Thus, rehab can dramatically change the economics of some energy conservation measures.

Shell measures such as insulation upgrades are often very costly, if not impossible, as part of apartment building energy conservation retrofit programs. But during rehab, a better job of insulation might entail a relatively small added expense that makes insulation improvement very cost-effective. For example, Paul Knight* recounted the rehab work in a masonry building where existing plaster lath walls needed replacement. The original rehab specifications

* Much of the discussion in this case study is based on phone interviews by John DeCicco with Paul Knight of Domus Plus, Oak Park, Illinois.

called for installing 2" x 2" metal channels and attaching drywall, but without insulation or air sealing. Knight's approach was to use 2" x 4" framing with cavity insulation and to carefully install the drywall so that it is effective as an air barrier. This technique will generally result in a cost-effective efficiency upgrade. A point to keep in mind when planning a rehab is that the renovated building is still affected by components that remain from the original structure—for example, when installing new interior walls with tight vapor barriers, one must consider the airflow and moisture dynamics of an old outer wall.

A program for energy-efficient rehabilitation was started by the Illinois Department of Energy and Natural Resources in 1988, motivated by the need to lower energy bills as a way to help keep rehabilitated housing affordable. The program provides grants of up to $2,000 per dwelling unit to cover the incremental costs of upgrading a rehabilitation to "superinsulated" standards. (Another part of the program is performing superinsulated rehabilitation of affordable single-family houses.) Base rehabilitation costs in these Chicago apartment buildings range from $50,000 to $80,000 per unit (Domus Plus 1994). According to Paul Knight, as of spring 1995, work had been completed on 23 buildings containing 408 units; another 10 buildings were in progress. Katrakis et al. (1994) analyzed preliminary performance data comparing conventional apartment building rehabs with superinsulated rehabs completed under the program. Control buildings receiving moderate rehabs had an average weather-adjusted space heating energy use index of 19 Btu/(ft^2·DD) (Btu/ft^2/degree day); superinsulated rehab buildings averaged 7 Btu/(ft^2·DD). The annual gas bill savings, estimated to average $355 for a 1,100-square-foot apartment, would cover one month of affordable housing costs for a household at the $14,000-per-year income level.

Ventilation and Air Leakage

In all buildings, the purpose of addressing air leakage and infiltration is not to minimize air flows, but rather to control them. Natural ventilation is not reliable for ensuring good air quality, which depends on how indoor spaces interact with outdoor temperature and wind conditions as well as on internal pollutant sources. For example, negative basement air pressure can draw radon or other soil gases into buildings; this effect can be quite strong in tall buildings. In apartment buildings, ventilation control means assuring adequate fresh air in the living spaces while avoiding cross-contamination and minimizing uncontrolled infiltration, which wastes energy and worsens mechanical system balance and control problems. The ASHRAE (1989) residential ventilation standards recommend outdoor air requirements of 0.35 air changes per hour (ACH), but not less than 15 cubic feet per minute

(CFM) per person. In a four-person, 1,000-square-foot apartment, for example, the per-person guideline would apply, yielding a recommendation of 0.45 ACH.

Relatively few data are available about air leakage characteristics in apartment buildings. Those data that are available must be used with caution and cannot be generalized to other buildings unless they are of very similar construction in similar regions. Accurate measurement of overall air leakage from large apartment buildings is possible only by tracer-gas techniques. Multiple blower door measurements in conjunction with modeling can also be used to estimate leakage rates (Diamond et al. 1986). The DOE continues to support research on understanding the interactions between air leakage and ventilation in apartment buildings, particularly in high-rise buildings, where mechanical ventilation is often an added complication to understanding how best to retrofit these buildings while still ensuring acceptable levels of ventilation. Lawrence Berkeley Laboratory (LBL) and CCC have been working on pilot projects in the San Francisco and Boston areas to improve ventilation and air leakage diagnostics and to develop retrofit guidelines for high-rise apartment buildings.

For practical retrofit purposes, it may be sufficient to identify bypass routes and apply blower door techniques if sealing is necessary in individual apartments. In apartment buildings, however, one cannot presume that excessive infiltration is a problem—it may be just the reverse. In fact, a building may be leaky on average—for example, from open windows due to uneven heating—whereas individual units may be underventilated because some tenants keep their windows tightly shut.

Some apartment buildings have been found to be quite leaky, with interunit flows being as important as flows to the outside. In a study of 11 apartment buildings in upstate New York, Synertech (1987) found an average pre-retrofit leakage rate of 35.5 ACH at 50 pascals (Pa). These rates are higher than those for both mobile homes and the top third of single-family dwellings, implying natural ventilation rates of over 1 ACH. Cameron (1990) reported 50-Pa leakage rates of 55 ACH in Philadelphia row houses, with up to 30% of the flow being across party walls (to adjoining units). Commoner and Rodberg (1986) examined three low-income apartment buildings in New York City using tracer-gas methods and found natural air leakage rates of 1.08, 0.58, and 1.01 ACH, about twice the leakage estimated from dimensional ("crack length") calculations. The researchers determined that the additional leakage was to the interior common spaces of the building and to neighboring apartments rather than through windows and exterior walls. This observation is consistent with the experience of a number

Air infiltration barrier being installed at the Northgate Apartments, a publicly assisted, privately owned complex in Burlington, Vermont.

of practitioners who have addressed similar building stock. Extensive air leakage has been found in recently constructed town house–type apartment buildings, where blower door and infrared thermography has also confirmed that air exchange among apartments is greater than air exchange with the outdoors (Fitzgerald 1990).

On the other hand, apartment buildings of more recent construction may actually be too tight. This situation indicates the need to be wary of air quality problems resulting from energy conservation efforts performed without adequate measurement-based diagnosis of building airflow characteristics. Baylon and Heller (1988) tested units in nine small, modern, motel-style apartment buildings (where each unit has its own outside door) and reported estimated average leakage rates ranging from 0.08 to 0.30 (median 0.19) ACH. Francisco and Palmiter (1994) tested three electrically heated apartment buildings, recently built to energy-efficient specifications in the Northwest. All three buildings were found to be seriously underventilated when natural infiltration was the only source of outdoor air; all units tested at below 0.31 ACH when standards called for minimum ventilation rates of 0.35 ACH to 0.42 ACH. Even when ventilation fans were running, some units failed to meet the requirements. From the results of this testing, the authors recommended updating energy efficiency guidelines to require continuous operation of apartment ventilation fans of adequate installed air-movement capacity.

Katrakis (1990) found that inadequately controlled infiltration in first-floor apartments could reduce energy savings in weatherization retrofits of walkup apartments. Three Chicago buildings were studied, all of which received mechanical system improvements. Cost-effective savings were achieved in two buildings. In the third building, adding a six-point averaging thermostat to the central single-pipe steam boiler controls and using vent treatments to better balance heat distribution yielded 13% space heating savings. However, average indoor temperature stayed very high (82°F), and the range between the coolest and warmest apartments rose from 10°F to 21°F. Tenants in lower apartments remained uncomfortable, and use of gas and electricity for supplemental heating rose, resulting in no net savings.

Apartment buildings can house numerous air leakage and bypass sites (see Table 3-2). In apartment buildings, air enters the lower levels through basements, around main entry doors, and through apartment windows. Air then escapes into plumbing or electrical shafts and other framing cavities, such as furred-out plaster walls, and through apartment doors into hallways, from which it can rise up stairwells, elevators, and other vertical shafts. At higher levels in the building, the air escapes through apartment or stairwell windows and through various rooftop openings. Potential convective paths in apartment buildings are listed in Table 3-2. Such air flow patterns, driven by the stack effect, can create comfort problems, such as cold lower-level apartments, and affect the heating balance for central systems (Katrakis 1990). Even in a

Table 3-2
Potential Convective Bypass Sites in Apartment Buildings

Common stairwells	Windows in stairwells	Doors at top of stairwells
Skylights	Elevator shafts	Attic scuttles
Kneewall access doors	Recessed light fixtures	Flush-mount light fixtures
Exhaust fans	Plumbing penetrations	Balloon-frame walls
Hollow masonry walls	Behind plaster lath	Behind paneling
Furred-out wall sections	Chimney clearance	Duct runs
Plumbing chases	Vent stack chases	Laundry chutes
Dumbwaiters	Unused chimneys	Exterior side walls
Interior partition walls	Beneath staircases	Dropped ceilings
Floor-joist cavities	Baseboards	Ceiling cove moldings
Door and window moldings	Rodent holes and tunnels	Built-in drawers and closets
Plumbing access panels	Electrical fuse boxes	Light switches
Electrical outlets	Holes for electrical wires	

two-story apartment building, researchers have found that ground-floor apartments can have heating bills more than double those of upstairs apartments because of internal air-leakage patterns and heat flow (McBride et al. 1990).

Problems with uncontrolled air flow can be particularly serious in taller buildings, where sealing up the first floor may simply move a cold-air infiltration problem up to the second or third floor. In such buildings, many services are run in a vertical path from the basement, often up to roof-level fixtures. Those who install and repair these systems are notorious for disregarding the need to maintain thermal envelope integrity. The taller the shaft and the greater the difference in temperature between inside and out, the greater the stack effect and the volume of air movement. Such problems can occur in any kind of structure, and newer buildings are not immune.

Experience in single-family dwellings shows that trying to stop infiltration at entry points alone can prove frustrating. Poor results with general weatherstripping and exterior caulking have led conservation specialists to identify key leakage sites from the interior and then selectively caulk, seal attic bypasses, and fill interstitial cavities with high-density cellulose (Fitzgerald 1989; Fitzgerald et al. 1990). Using blower doors and bypass sealing, Synertech (1987) documented reduction in leakage in a sample of apartment buildings by an average of 32%. These buildings were typically balloon-framed wood structures, having retrofit opportunities similar to those in large single-family dwellings. To address interior leakage paths, retrofitters can weatherstrip and assure proper closure of apartment doors, air-seal fire doors between floors, and use diagnostics to identify and seal the various chases that allow the air flow within the structure. This strategy is likely to apply in wood-framed or town house–style apartment buildings. However, for larger masonry structures, an intensive air-sealing strategy is unproven and could prove structurally damaging because of moisture migration.

Although windows often get attention because they are frequently the predominant perceived sources of cold air, they may not be the most important air leakage sites, even in larger apartment buildings. As noted earlier, window treatments can make sense as part of a well-thought-out retrofit package but are rarely sufficient alone, particularly when the major air leakage paths are to hallways, adjoining apartments, shafts, and other interior bypasses. Sealing such interior air leaks requires good diagnostic skills as well as ingenuity. It also entails utilizing a variety of materials, including urethane foams, flashing, or other barriers (such as custom covers and hatches). Such work must always be done with due attention to applicable fire codes, appliance venting requirements, the provision of sufficient dwelling space ventilation, and the avoidance of attic moisture problems, among other requirements. In particular, caution is needed when sealing interior leaks in buildings having central-corridor ventilation make-up for bathroom and kitchen venting. Caution is also needed in smaller structures with forced-hot-air systems using the hallways or stairs for return air. In both cases, either the free flow of air between common spaces and apartments should be maintained or alternate system modifications should be undertaken to assure adequate heat distribution and maintenance of satisfactory air quality. Clear understanding of what is going on in the system design and operation is critical, with due regard for maintaining adequate air quality.

CASE STUDY
Seattle City Light Retrofit Experience

The energy conservation programs of Seattle City Light (SCL) have provided extensive experience in retrofitting electrically heated apartment buildings. Beginning in 1986, SCL started retrofitting numerous buildings and has subsequently published evaluation results for various types of retrofits as well as the overall program (see Okumo Tachibana 1993). In the SCL service area, typical of the Pacific Northwest and others areas served by historically inexpensive and heavily subsidized hydropower, electric-resistance heat and hot water are common in apartment buildings. The SCL multifamily program focused on low-rise (up to four-story) apartment buildings having five or more dwelling units. A comprehensive set of relevant shell, water heating, and lighting retrofits was applied. The program was administered differently depending on the tenant income levels in the buildings. Low-income buildings received retrofit grants, at a total cost for the conservation measures of $2,343 per dwelling unit ($1,803 for installed equipment plus $540 for administration); standard-income buildings received low-interest loans and had a total cost of $1,428 per dwelling unit ($1,174 for equipment plus $254 for administration) (in 1990$, SCL 1991b). Electricity savings (as a percentage of pre-retrofit use) were measured separately for common-area (house) meters and unit (tenant) meters.

SCL evaluated 95 buildings with 1,365 units covered by the program in 1986–1987. Average annual electricity savings of 1,640 kWh per dwelling unit were obtained overall, with a similar absolute savings level in both low-income and standard-income buildings (Okumo 1991, 1992; SCL 1991a). These savings amounted to 4–9% (of 10–13 MWh/unit/yr pre-retrofit consumption) in low-income buildings and 13–18% (of 8–9 MWh/unit/yr pre-retrofit consumption) in standard-income buildings. At the relatively low power costs in the Northwest (a \$0.041/kWh avoided cost for SCL), the implied simple payback periods are quite long, 21 years in standard-income buildings and 34 years in low-income buildings. The 1986–1987 retrofit program was found to be cost-effective to the utility (compared with new electricity supply resources) only for standard-income buildings. Higher costs of measure installation and administration made the program appear not cost-effective in low-income buildings. However, interpreting evaluations of programs such as the SCL efforts can be challenging, particularly for low-income buildings with frequent tenant turnover, varying vacancy rates, and tenants of differing backgrounds and expectations with regard to energy use and comfort.

Window retrofits were found to account for the largest portion of savings; low-flow showerheads and common-area lighting retrofits also showed significant savings. Available window technology improved over the study period, from air-filled aluminum frame (U = 0.72) to air-filled vinyl frame (U = 0.62), so SCL estimates a higher energy savings for more recent window retrofits, using engineering calculations (measuring differences in achieved savings would be difficult and has not been reported). Thus, use of improved window technology is expected to help move the program toward cost-effectiveness (Okumo Tachibana 1993).

Recent years have brought increased activity to improve the efficiency of lighting and appliances in apartment buildings as part of utility conservation programs. Seattle City Light instituted a Multifamily Common Area Lighting Program in 1991, offering rebates for energy-efficient lighting retrofits in halls, utility rooms, parking lots, and other common areas of residential apartment buildings and condominiums. An evaluation of 11 buildings revealed electricity savings amounting to 50% of pre-retrofit lighting consumption or 11% of total house-meter consumption (Humburgs 1993). Lighting retrofits are discussed further under Lighting at the end of this chapter.

Domestic Hot Water

After space heating, domestic hot water (DHW) represents the largest energy end use in apartment buildings, accounting for approximately 30% of total energy use. Approximately 51% of apartment building DHW systems are gas fueled, with 38% electric resistance and the remaining 11% operating on fuel oil (RECS 1990a).

System Types

The wide variety of DHW systems found in apartment buildings implies a corresponding variety of conservation opportunities. Common types are individual tank heaters, central stand-alone heaters, tankless coils, indirect water heaters, separate boilers with storage tanks, and point-of-use heaters. A measurement-based characterization of system types and DHW usage patterns is not available for the United States as a whole.

Individual Tank Heaters

Individual stand alone tank heaters, similar to those installed in single-family residences, are often found in apartment buildings where tenants pay their own energy bills. These tank units are commonly electric heaters, installed in closets in individual apartments. In row housing or other low-rise buildings, gas- or oil-fired heaters are often installed in the basement (for appropriate venting) with pipes running up to the units.

Because these types of DWH heaters are typically purchased by the landlord or developer, but their operating costs are paid for by the tenant, purchasers have little incentive to select high-efficiency units or to install additional water heater insulation, pipe insulation, or other energy efficiency devices. When the heater is located in a closet in the living space, heaters are often undersized (20 gallons or less), and tenants may set tank temperatures very high to meet their needs, thereby increasing standby losses (and increasing air conditioning loads).

Central Stand-Alone Heaters

Large, central DHW heaters are often found in older buildings, particularly where tenants do not pay their own energy bills. One or two large units, most commonly gas- or oil-fired, are located in the basement of the building and supply multiple living units. Sometimes, separate units supply central laundry facilities and maintenance closets.

Due to the inherently intermittent demand on these types of systems, standby losses can be significant. Research by Ontario Hydro indicates that standby losses can account for up to 13% of total DHW energy use in apartment buildings (Perlman and Milligan 1988). Over- and undersized centralized units are common, particularly in buildings with variable occupancy or high turnover, which are often found in the low-income sector, and in moderate- and higher-income buildings, where tenants may install their own clothes washers and dishwashers. Goldner (1992) analyzed heating and DHW systems in 30 New York City apartment buildings and found that central stand-alone systems were typically undersized by as much as 25%.

Because of the often-extensive distribution system in apartment buildings, DHW tank temperatures are typically set much higher than for individual heaters (in excess of 140°F) so that occupants at the far end of the distribution loop receive hot-enough water. However, apartments at the beginning of the loop will then get scalding-hot water. The very hot DHW temperatures also exacerbate standby and distribution system losses.

Tankless Coils

In tankless coil systems, a heat exchanger coil is run inside (or just beside) the space-heating boiler to provide domestic hot-water heating. These systems are common in older, larger buildings. Seen as providing "free" hot-water heating in the winter (particularly when tied to an oversized boiler), they tend to operate at very low efficiencies; the water is commonly heated to a higher-than-desired temperature (particularly when used with steam boilers) and must then be mixed with incoming cold water to lower its temperature. Additionally, tankless coils are inherently very inefficient in the summer, as they require operation of the space heating boiler for a proportionally very small load, typically 10% or less of space-heating load (Sachi et al. 1989).

Indirect Water Heaters

Indirect water heaters operate similarly to tankless coils in that the source of DHW heating is the space boiler. However, in these systems the heat exchanger is used to heat water in a separate storage tank, rather than supply DHW directly. The efficiency of these systems depends on the efficiency of their components; when used in tandem with a high-efficiency modulating or condensing boiler, indirect water heaters can be highly efficient.

Separate Boilers with Storage Tanks

Here, a separate, non-space-heating boiler provides domestic hot water to a storage tank. Systems having a separate boiler and storage tank are typically found in older large buildings. They operate similarly to large central stand-alone heaters; however, they incur heat losses associated with two pieces of equipment (the boiler and storage tank) instead of one, plus losses associated with the transfer of water from the boiler to the storage tank. For this reason, these systems may operate at relatively low efficiencies; but, again, system efficiency depends on how well system losses are minimized in the design, installation, and maintenance of the components.

Point-of-Use Heaters

Point-of-use heaters are small units typically found under or above a sink to provide DHW for only that faucet. Commonly found in commercial applications, they are sometimes installed in apartment buildings, particularly high-rise, upper-income buildings. These units are sometimes equipped with a small amount of storage capacity (for example, less than 5 gallons) or are simply resistance coils that heat water on demand. They eliminate the need for distribution piping and thus eliminate the distribution system losses associated with other types of DHW systems. However, units with storage capacity in low-use areas may have significant standby losses. If used in appropriate applications, these units can be highly efficient.

Distribution Systems

With the exception of point-of-use systems, all DHW systems require some sort of distribution system. Individual-unit stand-alone heaters generally have limited distribution piping, leading from the heater only to the bathroom and kitchen in a living unit. Because piping from individual water heaters largely runs through conditioned spaces and because water is distributed through the system only on demand, heat loss through distribution piping is typically minimal. However, in centralized DHW systems, distribution piping configurations can be quite extensive. If the distribution system is poorly designed, piping is uninsulated, or piping goes through unconditioned spaces, substantial distribution losses can occur. With large systems, often a small pump (1/4 to 3/4 horsepower) distributes hot water throughout the system at all times, eliminating long waits for hot water at a faucet. Continual circulation of hot water, particularly at times when there is very little demand, can further increase piping heat losses. DeCicco (1988b) found that distribution losses accounted for 30% of seasonal water heating energy use in a 60-unit apartment complex with a central system and an uninsulated circulation loop.

Scale buildup, leaks, deteriorated insulation, and so on, resulting from poor distribution system maintenance, all can add to the energy demands of the DHW system. Instrumented studies of apartment building DHW systems have revealed some relatively high leakage rates. A study of an apartment complex in California found leaks accounting for 20–30% of building DHW usage (Vine et al. 1987). In a New Jersey apartment complex with a central system, measurements showed that 19% of total hot-water usage was due to system leaks (DeCicco 1988b). In one Minnesota apartment building, measured DHW leakage was 23% and was characterized by the researchers as "disturbingly high since the maintenance in this building is above average" (Sachi et al. 1989).

Energy Conservation Opportunities

Lowering Temperatures

The most common recommendation for hot-water energy conservation in most residential settings is the reduction of DHW temperatures. Many stand-alone heaters come factory-set at 140°F, which is higher than necessary for most residential services and poses a risk of scalding. ASHRAE (1991) recommends temperatures from 105°F to 115°F for lavatories and showers. In central systems, antitampering devices may be warranted to keep tenants from adjusting DHW heater aquastats. Although dishwasher temperature needs are relatively high, most dishwashers are equipped with booster heaters that can elevate temperatures for the 8 to 14 gallons of water required for each load. Most booster heaters are designed for 120°F incoming water and boost temperatures to 140–145°F (Wilson and Morrill 1995).

In apartment buildings with individual tank heaters, a separate DHW heater is often used for common laundry facilities; this water heater is typically set at a higher temperature than that of individual apartment heaters. Recent studies, however, show that 140°F incoming hot-water temperatures offer little improvement in clothes washer performance as compared with 120°F incoming temperatures (Wilson and Morrill 1995). Thus, where individual-unit clothes washers are in place, it may be possible to leave the individual apartment water heater at 120°F to meet all service needs.

For systems with circulation loops, DHW setpoint temperatures can be reduced during periods of low use. Lobenstein, Bohac, Korbel, et al. (1992) analyzed the savings associated with DHW temperature reduction through control of circulation-loop temperatures during periods of low demand. Three apartment buildings were studied, two with 39 units each and one with 47 units. Each building was equipped with two centralized gas-fired stand-alone heaters rated at 200,000–250,000 Btu/hr, and the buildings had annual DHW operating costs ranging from $3,716 to $5,631. Each building had two types of electronic controls installed: a timer that set back temperatures automatically from 140°F to 115°F at preset times during weekdays and weekends, and a demand-based controller that modulated the temperature between 115°F and 145°F according to burner firing time (that is, increased firing meant higher demand and the need for higher temperatures). For rotating one-week intervals over one year, the heaters were monitored operating in each of the three modes: with the existing aquastat set at 140°F, with the time controller operating, and with the demand controller operating. Savings for the time control averaged 10.3%, whereas savings for the demand controller averaged 16.2%. Seasonal efficiency improvements were estimated at 3% for both control strategies. With installed costs of $1,000 for the time control and $1,400 for the demand control, the devices had simple paybacks of 2.2 and 1.9 years, respectively.

Care should be taken not to set DHW temperatures too low for long periods of time, however, because of concerns regarding Legionnaire's disease, a severe respiratory infection identified with an outbreak during a 1976 American Legion convention. The infection is caused by *Legionella pneumophila* bacteria, which are transmitted mainly through inhalation. These bacteria have been found in water samples from homes with electric water heaters. Several outbreaks in the United States were traced to *Legionella pneumophila* growth in showerheads; however, these were all found in large commercial applications (that is, hotels and hospitals). *Legionella* bacteria can colonize in DHW systems maintained at 115° or lower (Ciesielki 1984, as cited in ASHRAE 1991). An Electric Power Research Institute (EPRI) epidemiologic study found that use of electric water heaters was not a statistically significant risk factor in homes of Legionnaire's cases. Although use of electric water heaters was more common in those cases, it was highly correlated with having a nonmunicipal water supply, which, along with smoking and having had recent plumbing work done, were the three factors found to significantly explain disease incidence (EPRI 1995). There are no specific guidelines regarding this issue for apartment buildings, making it an area for further study.

Another potential energy-saving measure for individual apartment heaters involves shutting off or turning down units during unoccupied periods. This measure is particularly appropriate for electric heaters accessibly located within individual apartments. Some DHW heaters are equipped with a "vacation" setting on the aquastat to facilitate this process. However, tenant education is key for this measure to be implemented.

Tank Insulation

Although newer stand-alone hot-water heaters may have adequate insulation in the tank jacket, additional insulation wraps can still be cost-effective, especially if units are located in unconditioned spaces and for units that are undersized or that operate at high temperatures.

Vent Dampers

Several studies have been performed to determine the cost-effectiveness of vent (flue) dampers on fuel-fired water heaters. As a retrofit, dampers must be installed downstream of (that is, above) the draft diverter in the water heater's flue. Factory-installed (integral) flue dampers, on the other hand, are located upstream of (below) the draft diverter, and so are better at retaining heat inside the water heater. Lobenstein, Bohac, Staller, et al. (1992) found that in apartment building applications, large commercial tank heaters with integral flue dampers had efficiencies ranging from 56% to 67%, whereas those without integral flue dampers had efficiencies ranging from 53% to 64%. They found that thermally activated dampers generally cannot achieve a good seal, with tests on water heaters failing to show reliable, cost-effective savings. But electronic dampers have relatively high costs as an add-on measure, and so they are unlikely to be cost-effective as a water heater retrofit. With a $540–$640 installation cost, Lobenstein, Bohac, Korbel, et al. (1992) and Lobenstein, Bohac, Staller, et al. (1992) estimated paybacks ranging from 6 to 30 years, which are longer than desirable for most apartment building owners. However, as noted earlier in the section on boiler retrofits, it may be desirable to add a damper downstream from the draft diverter of a DHW heater that shares a flue with a boiler. Because of the new efficiency standards for commercial tank-type water heaters, most such heaters manufactured since 1994 will include integral flue dampers.

System Replacement

Since DHW systems in apartment buildings are often oversized, downsizing plus installing higher-efficiency equipment can potentially yield large energy savings. A study monitoring 30 apartment buildings in New York found that combination DHW heating systems were oversized by 20–300% (Goldner 1992). However, there is great variability among operating efficiencies of apartment building DHW systems, and careful analysis is needed to justify DHW heater replacement. Experience has shown that whole-system replacement can rarely be justified by energy savings alone. In two small (under ten-unit) buildings, Lobenstein, Bohac, Korbel, et al. (1992) tested a high-efficiency condensing water heater as a replacement for a conventional stand-alone gas-fired DHW heater. Annual savings were large, estimated at 28% for both of the test sites, but payback times were quite long, 24 years and 28 years, respectively—longer than most building owners would consider worthwhile.

System replacement with modular front-end boilers is commonly considered for older, large, combination space and DHW heating boilers. In these cases, it is likely that a front-end boiler will produce domestic hot water more efficiently than a large combination boiler because of reduced standby losses and higher combustion efficiencies. One particular opportunity is in buildings where large space heating boilers run year-round to provide hot water in summer months; the potential savings may be large enough to justify investing in smaller efficient boilers to replace the large boiler during summer operation. The case study of a St. Paul high-rise (see Case Study: Hot-Water System and Lighting Retrofits in a High-Rise Public Housing Complex, below) includes such a retrofit. Several other apartment building retrofit projects have shown cost-effective retrofits of this type (DeCicco and Dutt 1986; Robinson et al. 1988). However, a number of studies have shown only modest or no savings, as the higher efficiency of the combination system during winter operation offsets the lower efficiency during summer operation, thus rendering annual performance similar to that of a front-end boiler (Englander and Dutt 1986; Sachi et al. 1989).

Pipe Insulation

Bare copper pipe has two to three times the heat-loss rate of the same size piping with 1/2" fiberglass insulation; potential energy savings from pipe insulation are even greater for pipe runs through unconditioned spaces (ASHRAE 1991). However, adding insulation in existing buildings is often difficult because of the inaccessibility of much of the distribution piping runs (which are contained in walls, crawl space, and so forth) and therefore may not be cost-effective for the small amount of piping that is accessible.

Circulation Pump Control

In large centralized systems, a small circulation pump frequently runs 24 hours per day to continuously circulate hot water throughout the building. However, during periods of very low demand, it may be economical to shut off the pump. This measure can greatly reduce distribution system losses and also afford a small amount of electricity savings for the pump itself. Hot water would still be available to tenants during these off-cycle periods but would require a longer wait at the tap. As discussed earlier, an alternative strategy is reducing the DHW setpoint temperature during periods of low demand.

Low-Flow Devices

Reducing both water and energy waste by using less wasteful plumbing fixtures is a well-known and effective conservation strategy. The Energy Policy Act of 1992 requires that all showerheads, bathroom and kitchen faucets, and replacement bathroom and kitchen aerators manufactured after January 1, 1994, have flow rates of 2.5 gallons per minute (gpm) or lower (at 80-pounds-per-in^2 water pressure) (EPACT 1992, Section I:C:123j). Previously, many showerheads and faucets with flow rates far exceeding this were sold. Commercially available faucet aerators with flow rates of 1.5 gpm or even lower are acceptable in many apartments.

Seattle City Light included low-flow showerheads in its Multifamily Conservation Program in 1986–1987. Okumo (1992) found that the showerheads accounted for about 10% of the 10–14% living-unit savings (with the remaining savings coming from lighting and HVAC measures). However, significant resistance by customers toward low-flow showerheads led to only a 41% saturation rate among participating households.

Similarly, the city of San Jose, California, found that showerhead replacement was very cost-effective in apartment buildings. However, program administrators learned that direct installation by the canvasser who delivered the retrofit device was essential, as opposed to leaving it with the occupant to install. This arrangement required advance scheduling with building management to provide access to living units. In contrast, single-family-dwelling retrofit could be successfully accomplished through several methods of delivery, including direct installation or occupant installation (Jordan 1990).

At the Eden Drive Apartments in Connecticut, CCC recommended the installation of pressure-reducing valves, mixing valves, low-flow faucet aerators, and low-flow showerheads in this 60-unit, all-electric public housing complex. Extremely high water pressure (measured at 90 psi) had caused faucet washers to prematurely deteriorate, resulting in leaks. Installation of pressure-reducing valves dropped pressures to 40–60 psi. Existing showerheads, measured at 2.5- to 3-gpm flow, were replaced with 2-gpm showerheads. Low-flow faucet aerators were installed to replace those that were missing or broken (because of the high-water-pressure debris collected in old aerators and damage from dishwasher and washing machine hookups). Annual DHW savings from these measures were estimated at $5,509, corresponding to 4% total building electricity savings and resulting in a 6.2-year simple payback. Metered whole-building savings could not be broken out by end use, but exceeded projections, indicating that the DHW measures were likely very cost-effective (CCC 1994; Kamalay 1992).

CASE STUDY

Hot-Water System and Lighting Retrofits in a High-Rise Public Housing Complex

In 1991, the Ramsey Action Program (RAP) of St. Paul, Minnesota, completed weatherization work on a high-rise apartment building located at 727 Front Street in St. Paul and owned by the city's Public Housing Agency.* Built circa 1970, the building has 151 units and 112,000 square feet of conditioned space. Two large 4.4-MBtu/hr boilers provide multizone two-pipe steam heat as well as domestic hot water (and thus had been operated year-round). Fuel (gas and oil) consumption records were poor for this building since the boilers are fired by interruptable gas with fuel oil backup and oil use records were not available. The Public Housing Agency estimated a pre-retrofit energy use index of 12.2 Btu/(ft$^2 \cdot$DD); this value implies an annual fuel consumption of 11,000 MBtu, costing $38,000 at $3.44/MBtu. Pre-retrofit annual electricity consumption was 2,953 MBtu (865 MWh), costing $36,700 at 4.3¢/kWh.

Almost all of the 727 Front Street retrofit work was with the central mechanical systems and common-area lighting. The project was completed in April 1991; total installed cost of the retrofits was $39,080. The only architectural measures or individual-apartment retrofits were the installation of room air conditioner covers. RAP planned lighting retrofits, separate two-stage modulated boilers for summer water heating, and various other central plant measures. However, not everything on this job went exactly as planned.

Two modular front-end boilers (240,000 Btu/hr output each) were installed for summer domestic water heating. Although RAP recommended that these boilers be run as staged units, building management chose to use the boilers for two separate zones and to run both units continuously from May 15 through September 15 each year. In addition, RAP insulated all exposed DHW pipes as well as the condensate return line from the space heating boiler (total cost, $900). Condensate had been returning at 180°F yet sometimes froze as it passed the louvered combustion air intakes for the boiler. This problem was also corrected by repairing the damaged combustion air dampers. RAP also recommended running the large boilers in the winter as staged units and installing turbulators (which improve heat-exchange efficiency) in the fire tubes. However, these two steps were not taken, so little work was done that would have a significant impact on space-heating fuel use.

The installed cost of the separate summer boilers was $10,130. Although heating-season fuel use records are unreliable, good sum-

---

* This case study is based on interviews with staff of the Ramsey Action Program, plus data collection and analysis conducted by Tom Wilson.

mer savings estimates were obtainable since there are no gas interruptions during the summer. Pre-retrofit summer gas use was 919 MBtu, and estimated savings were 245 MBtu (27%), valued at $844 per year. These estimates imply a simple payback of 12 years and a cost of conserved energy of $0.96/MBtu (assuming 20-year lifetime and 5% real discount rate but not adjusting for added maintenance costs of the new boilers). The benefit/cost ratio calculated according to WAP guidelines was 1.04. Thus, this summer water heating retrofit was cost-effective even though realized operation of the new system was suboptimal.

The common-area lighting upgrade was also successful in terms of cost-effectiveness. However, it turned out to be a painful disappointment for the local agency because the retrofitted costs were rejected by WAP administration. The complete change-out of common-area lighting cost

This high-rise public housing project on Front Street in St. Paul was retrofitted with modular boilers for summertime water heating and more efficient lighting in common areas.

$22,400, which included replacing incandescent fixtures with fluorescents in stairwells and exit signs. Existing common-area fluorescent fixtures were retrofit with new reflectors, and the old ballasts were replaced with high-frequency ballasts. The local utility, Northern States Power, provided a grant to pay for proper disposal of the old ballasts. Analysis showed an electricity savings of 170 MWh, 20% of pre-retrofit consumption, valued at $7,200 per year. The implied simple payback was 3 years, and the benefit/cost ratio was 4.01. Nonetheless, since at that time lighting retrofits were not yet an allowable WAP measure, program administrators refused to fund the DOE portion of the expenses and required RAP to pay the $22,400 back to the state. Nevertheless, this project demonstrates the typical good value of common-area lighting retrofits in apartment buildings. Since lights in common areas are usually on 12 to 24 hours per day, excellent savings are possible with a well-designed lighting retrofit.

Household Appliances

Reducing appliance energy use in apartment buildings can be a more difficult problem than it is in single-family dwellings. One reason is the split-incentives barrier, which inhibits purchase of efficient appliances for new construction, replacement of inefficient equipment with higher-efficiency appliances in existing structures, or operation and maintenance of appliances and equipment for maximum efficiency. Another reason is the lack of efficient equipment targeted to this market, which often emphasizes smaller, low-frills appliances with a strong preference given to low first cost.

Laundry

Apartments are much less likely than single-family residences to have individual clothes washers and dryers (comparative saturation rates are given in Chapter 2). In many apartment buildings, washers and dryers are provided in common areas for shared use among building occupants. The most significant energy use associated with laundry is water heating—accounting for approximately 90% of energy use in clothes washing (Wilson and Morrill 1995). There can be up to a 20:1 difference in energy used for a hot-wash/hot-rinse cycle compared with a cold-wash/cold-rinse cycle, depending on the water heater temperature and type of appliance. Hot-wash/cold-rinse or warm-wash/warm-rinse cycles can cut energy use in half compared with a hot/hot cycle.

Higher washer spin speeds extract water from clothes more effectively, reducing subsequent energy use in drying, and horizontal-axis washers have much higher spin speeds than vertical-axis machines. Several major manufacturers are expected to introduce horizontal-axis machines into the U.S. market over the next few years, encouraged by the higher efficiency standards likely to be issued by the U.S. Department of Energy. The most important energy-saving feature in clothes dryers is moisture-sensing automatic shutoff, which avoids overdrying.

Although end-use hot-water loads generally account for a proportionately smaller share of overall water heating energy use in central systems compared with individual-unit water heating systems, reducing unnecessary hot-water consumption is still a cost-effective strategy. Resident education programs should encourage occupants to use hot water for clothes washing only when necessary and to wash or at least rinse in cold water whenever possible. Occupants should be encouraged to wash a full, large load whenever possible or, if a smaller load of clothes must be washed, to use a lower fill setting if available.

Dishwashers

Apartment households are slightly less likely to have dishwashers than are single-family residences, with 45% versus 50% penetration (RECS 1990a). Apartment dishwashers may be compact-sized machines (holding less than eight place settings). Approximately 80% of the energy associated with automatic dishwashing is associated with hot-water heating (Wilson and Morrill 1995). Automatic dishwashing can use less hot water (and thus less energy) than washing by hand if (1) hot-water temperatures are set at proper levels and (2) dishwashers are run with full loads. It is also important to select the proper wash setting (light soil, regular, or pot scrubber, for example). A dishwasher set on the "pot scrubber" setting can use almost twice the water as when set on "regular." Also, selecting the "energy-saving dryer" feature will reduce energy use by circulating room air through the machine to dry dishes rather than using the electric-resistance heater within the dishwasher.

Refrigerators and Freezers

Refrigerators and freezers account for approximately 8% of electric energy use in apartment units. Although federal standards for refrigerators have been increasing the minimum efficiency levels of new refrigerators, appliances in apartment buildings are typically older than those in single-family residences. About 47% of refrigerators in apartment buildings have manual defrost, compared with 20% of units in single-family residences (RECS 1990a). Manual defrost is more common in smaller as well as older refrigerators. Older frost-free refrigerators (dating from ten or more years ago) are often very inefficient; older manual-defrost refrigerators are not as bad.

Refrigerators of greatly improved efficiencies are now entering the market, thanks to the national appliance-efficiency standards and, more recently, the Super-Efficient Refrigerator Program (L'Ecuyer et al. 1992; Feist et al. 1994). The new "super-efficient" refrigerator is a large (22-cubic-foot), feature-laden, side-by-side model that would not be appropriate for most apartment building settings. This large refrigerator has an average-use electricity consumption rating of 670 kWh/yr, just above the top of the 530–660 kWh/yr range for new apartment-sized (14- to 16-cubic-foot) units in 1994. Although not "super-efficient," today's commonly available new units are still a substantial improvement over those from a decade ago, which ranged from 840 to 1,270 kWh/yr for apartment-sized refrigerators (Morrill 1995).

Installation of the most efficient models available should be encouraged during major renovation and in new apartment building construction. HUD guidelines should be amended to allow Public Housing Authorities to purchase "super-efficient" refrigerators under the Performance Funding System (as described in ORNL 1992). Similar arrangements already exist for efficient heating and hot-water equipment.

Currently, the federal government subsidizes the purchase of roughly 100,000 refrigerators annually and also subsidizes the energy bills resulting from their use; it is therefore in the long-term financial interest of HUD for housing authorities to obtain the most energy-efficient appliances on the market. Although "super-efficient" refrigerators may not be available in apartment sizes until they are required by standards, their availability could be accelerated through a federally coordinated purchase aggregation program. A valuable model for such efforts is the bulk purchase initiative recently started in New York for bringing an apartment-sized super-efficient refrigerator to market. Involving CEE, DOE, the New York Power Authority (NYPA), and the New York City Housing Authority (NYCHA), this program will involve aggregated purchase of a 14-cubic-foot refrigerator to be developed by manufacturers with both higher efficiency and a timer for avoiding the defrost cycle during peak power demand periods. The initial arrangement involves NYPA acting as purchaser for NYCHA, which provides a model for other potential third-party purchase and shared-savings arrangements involving refrigerators and other appliances. We make recommendations along these lines in Chapter 6.

Cooking

Stove and oven use in apartment units is typically a very small fraction of total living-unit energy use and generally resembles single-family household use (RECS 1990a). A few studies monitoring apartment-unit cooking gas use have been reported. McBride et al. (1990) related seasonal cooking gas patterns in 39 small row house–type apartments of mixed occupancy in a low-income Washington, D.C., neighborhood: consumption averaged 20,000 Btu/apt/day, but among different apartments the range spanned from 4,000 Btu/apt/day to 46,000 Btu/apt/day; seasonal variability was from 13,000 Btu/apt/day June–August to 30,000 Btu/apt/day October–December. DeCicco (1988b) measured both hourly and seasonal profiles in a 60-unit seniors complex in New Jersey: the six-year study average was 24,000 Btu/apt/day, but with strong seasonal variability, ranging from lows of around 15,000 Btu/apt/day in August to peaks of 35,000 Btu/apt/day in October, with smaller peaks sometimes noted in May.

Both studies just cited also reported use of kitchen ranges for supplemental space heating. This practice is well known among conservation practitioners, particularly for poorly maintained buildings in which heating is inadequate or poorly controlled. A kitchen range is a costly and hazardous way to provide space heating. Tenants may find it tempting, however, where space heating is provided by a tenant-paid fuel and cooking is supplied by a landlord-paid fuel (for example, a gas stove on a central building meter in an electrically heated, individually metered apartment). With gas stoves, this practice can also be extremely dangerous, leading to elevated (and even potentially fatal) carbon monoxide (CO) concentrations and fire hazards. Hazardous CO concentrations from ovens and ranges have been found to be disturbingly common in low-income settings; Tsongas (1995) describes procedures for testing and tuning gas ovens.

Miscellaneous Appliances

Saturation levels and use of smaller appliances (microwave ovens, televisions, and so on) are fairly similar for households in single-family and apartment buildings, with slightly lower saturations in apartment households for newer, more expensive devices (such as videocassette recorders). Appliances in apartments are likely to be older and more poorly maintained than those in single-family residences and therefore often operate at lower efficiencies (RECS 1990a). However, since their contribution to energy use in apartment buildings is fairly small and since efficiency improvements might best be obtained by programs directed at manufacturers, we are unaware of any particular need to address energy use by these appliances in the multifamily sector.

Lighting

Lighting usage within apartment units is similar to that in single-family dwellings on a per-square-foot basis. However, the additional energy usage from common-area lighting and outdoor lighting in apartment buildings (which is expended at much greater densities and for longer hours than in single-family residences) increases the total energy usage attributable to lighting above that of single-family dwellings (RECS 1990a).

As with single-family residences, the greatest opportunity for lighting efficiency improvements is in converting from incandescent lighting to fluorescent. According to RECS (1990a), 6% of apartment households reportedly have one fluorescent light that is left on for more than four hours per day, as compared with 9% of single-family dwellings. Upgrading to fluorescent lighting can be accomplished through replacement of existing fixtures with new fixtures for conventional straight-tube, "Circline," or compact fluorescent lamps (CFLs). CFLs that can fit into many existing incandescent fixtures are also available. Fluorescent lighting is four to five times more efficient than incandescent—that is, the light output (measured in lumens) per watt is four to five times greater for fluorescents than for incandescents. Moreover, fluorescent lamps are typically rated at 8,000- to 12,000-hour lives, whereas standard incandescent lamps have only 1,000- to 1,500-hour lives. Fewer lamp changes are needed with fluorescent lighting, resulting in lower maintenance costs and much lower lifecycle costs overall.

Electric utilities across the country offering rebates, coupons, and other incentive programs to increase the market penetration of CFLs in the residential sector have met with mixed success. CFLs are often too large or too heavy to fit into many existing incandescent fixtures, especially table and floor lamps. Another problem is that CFLs operate best in the "base-up" position, whereas residential screw-base fixtures generally require "base-down" operation. Studies at the Lighting Research Center (Troy, N.Y.) have found that CFLs operated "base-down" have anywhere from 5% to over 20% lumen depreciation as compared with "base-up" operation (Davis 1993). New England Electric found that approximately 25% of CFLs were taken out and replaced with standard incandescent bulbs within the first several months of installation because of inadequate light output, discomfort with light color, problems with lamp flickering upon startup, and other aesthetic issues (Jacobson et al. 1992). These problems are being overcome as manufacturers introduce smaller, lighter, higher-output, and flicker-free CFLs with better light color qualities. CCC's retrofit of the Eden Drive Apartments included replacement of two-lamp incandescent fixtures in unit bathrooms and kitchens with two-lamp CFL fixtures. The original fixtures were operating with either 60-W or 75-W incandescent lamps and were replaced with 13-W compact fluorescent lamps. Savings of $2,066/yr were projected with an investment of $12,000, resulting in a simple payback of 5.8 years (Kamalay 1992).

Common-area lighting is an important end use to consider in the multifamily sector (see the preceding case study, which included lighting retrofits in a Minnesota high rise). Lights in hallways, stairwells, elevators, lobbies, utility rooms, and other common areas are frequently left on 24 hours per day. High-efficiency fluorescent lighting (and perhaps motion sensors) in utility, laundry, and storage rooms can reduce lighting costs. Okumo (1992) found house-meter savings of 21% for low-income buildings and 36% for standard-income buildings from replacement of incandescent lamps with CFLs in apartment building common areas in Seattle City Light service territory. Later monitoring of 11 buildings participating in this project revealed savings of approximately 50% of pre-retrofit lighting consumption, or 11% of total house-meter consumption (Humburgs 1993). The lower savings found in the later study may be a coincidence of the particular subset of retrofitted buildings chosen for the later study or may indicate some problems with persistence of savings for these measures.

Similarly, 17 buildings in Wisconsin monitored before and after retrofit of all common-area lighting in operation 24 hours per day showed 30–33% measured savings with a simple payback of about one year. These retrofits included incandescent fixture replacement with dedicated CFL fixtures; the project managers cited concerns with improper fit of screw-base CFLs in old incandescent fixtures and the likelihood of theft of screw-base CFLs. They particularly noted the attractiveness of exit sign retrofits for apartment buildings, where two 30-W incandescent bulbs were replaced with one 9-W CFL (Hasterok 1990). Stum (1992) found ample potential for cost-effective use of fluorescent lighting in typical new apartment building common areas.

Due to problems with occupant acceptance of CFLs in existing fixtures, as demonstrated by the above case studies, it appears that common-area lighting is an appropriate first target for lighting retrofit in apartment buildings. Because common-area lighting fixtures are bought by the landlord and the costs of operating these fixtures are also paid by the landlord, this retrofit strategy avoids the split-incentives problem. Although lighting retrofit of existing incandescent fixtures with fluorescent lighting in individual apartments can be very cost-effective, tenant education and follow-up are important, since tenants are likely to remove the new bulbs if they have problems with them (Jacobson et al. 1992). Hasterok (1990) points out the value of replacing incandescent fixtures with dedicated fluorescent fixtures, where appropriate, to avoid problems with fit of screw-base CFLs and reversion to use of incandescent lamps. This approach can cut down on theft problems as well.

Although there is a dearth of literature on experience with exterior lighting retrofits in the multifamily sector, it is well established that lighting energy use can be greatly reduced with use of high-efficiency light sources, such as metal halide or high-pressure sodium. Maintenance savings can also be achieved with these light sources, as they have very long lives and thus reduce the number of lamp changes needed. As exterior fixtures are commonly left on all night, savings can also be achieved with use of photocells or timers. However, care should be taken to use tamper- and vandal-proof fixtures and controls to avoid costly equipment replacement or occupant overriding of controls.

Conclusion

Technical knowledge of how to effectively retrofit apartment buildings has undoubtedly improved over the past decade. This chapter has covered a variety of building types, but coverage is still lacking for some regions, particularly the Sunbelt. Our review was not able to cover new construction, which, where it is taking place, is generally of buildings different from most of those for which energy conservation experience exists. Even where experience does exist, uncertainties remain and confidence is still hampered by a lack of measured performance results. Recommendations for further research are presented in Chapter 6.

Nevertheless, the general principles that have been learned are likely to be transferable throughout this varied sector. One is the importance of the "building-as-a-system" approach, in which those seeking to improve the energy efficiency of a structure learn to think about the building as an integral whole and within its institutional context, including the important roles played by the people involved—the building owner or management, staff, and residents. Experience is the best guide to effective audit and retrofits; therefore, those approaching new types of buildings and regions must do so carefully, taking the time to learn—through measurements and observation—about the structure and its equipment, how it is operated, and who makes the decisions. Even though the largest per-unit energy savings are likely to be obtained with comprehensive approaches targeted to high-consuming buildings, there are still opportunities for selective, prescriptive approaches, such as some lighting and water conservation retrofits, which can be appropriate for utility programs, for example. Chapter 4, which describes program experience in apartment building energy conservation throughout the United States, turns to the topic of how to put into practice the technical knowledge just reviewed.

Programs for Apartment Building Retrofit

Not only have there been relatively few retrofit programs specifically designed for apartment buildings, but there have been even fewer evaluations of the success of these programs. The federal government did not treat apartment building energy efficiency issues as distinct from those of the single-family sector until 1979, when the U.S. Department of Energy commissioned a report, *Achieving Energy Conservation in Existing Apartment Buildings* (DOE 1985). Electric and gas utilities did not begin implementing energy efficiency programs specifically designed for apartment buildings until the mid-1980s. To date, most utilities still have no such programs.

The first section of this chapter, Program Experience, gives an overview of public, utility, and private-sector programs that address energy efficiency improvement in apartment buildings, tracing the history of residential efficiency programs and describing programs that specifically affect the multifamily sector. The second section, Program Evaluation, discusses methods used to evaluate apartment building energy efficiency programs and raises issues involved in such evaluations. We then present a selection of case studies of apartment building retrofit program evaluations.

Program Experience

Apartment buildings are the most underrepresented category of housing in government-funded and utility-sponsored energy efficiency programs. In 1981, a study addressing governmental roles in energy conservation for multifamily housing in the Northeast noted:

> The solution to the Northeast's energy crisis in multifamily housing is not likely to come from the federal level. Federal policies designed to encourage conservation and solar energy investment in multifamily housing have had a tendency to lump rental housing with owner-occupied housing in an effort to design energy policies and programs for the residential sector as a whole. As a result, incentives have been too small, too few, and of the type least attractive to rental housing owners. (Raab and Levine 1981, 62)

The authors went on to encourage state and local governments to implement "effective multifamily building energy conservation strategies" because such entities have greater flexibility to respond to the diversity in local housing markets. The situation today in the Northeast as well as in other parts of the country is not remarkably different from that described above. Federal programs for residential energy efficiency still remain largely centered on single-family and owner-occupied housing. State and local energy efficiency programs have not met their potential for specifically targeting the multifamily housing sector. Although some utilities have developed multifamily programs, these programs are few in number and limited in scope.

Federal Energy Conservation Programs

Although energy conservation is now often viewed as a proactive initiative to save both energy and financial resources, the first federal energy efficiency programs in the United States were reactive, part of the "crisis" response to the 1973 Organization of Petroleum Exporting Countries (OPEC) oil embargo. Low-income households were hit especially hard, and many were unable to cover the rapidly rising cost of household energy. The federal government responded to the energy crisis by authorizing fuel assistance and weatherization programs on behalf of low-income households. The Office of Economic Opportunity (what was then the federal antipoverty agency), and later the Community Services Administration, authorized local community action agencies to implement programs in both crisis assistance (payment of energy bills) and residential weatherization to reduce low-income households' energy cost burden.

Weatherization Programs

The first weatherization programs were intended to increase social equity via provision of services to low-income households in the face of rising energy costs. The Community Services Administration (CSA) program for weatherization was the first and only federal program of its kind when it was implemented in 1974, but other federal energy efficiency programs and initiatives followed. The CSA weatherization program authorized local community action agencies to use up to 10% of their CSA general operating grants to help low-income households achieve a greater degree of energy efficiency and also to help pay energy bills. The CSA grantees would provide, at no cost to the residents, an energy audit and the installation of energy conservation measures, up to a level of about $450 per household.

The Energy Conservation and Production Act of 1976 authorized the creation of a program devoted solely to energy conservation, the Low-Income Weatherization Assistance Program (WAP). Administrative responsibility for WAP was transferred from CSA to DOE, created by the Department of Energy Organization Act of 1977. Under WAP, the federal government provides funding to state weatherization agencies via DOE's regional offices. The states then contract with local community action agencies—or in some cases with local governments, other nonprofits, or Native American tribes—to provide weatherization services. About two-thirds of the 250,000 units weatherized every year by local weatherization agencies are treated under DOE WAP rules, limiting this program to households with incomes below 150% of poverty level and providing a maximum of $1,600 in conservation services per household. Although individual apartments with qualifying tenants may be treated separately, for an apartment building to be weatherized under WAP, at least 66% of the units in the building must meet the low-income requirements.

The Low Income Home Energy Assistance Program (LIHEAP), administered by the Department of Health and Human Services (HHS), also provides funds for weatherization. Begun in 1982 as a block grant to states, LIHEAP provides assistance to low-income households that are unable to pay their energy bills. Total LIHEAP appropriations were $1.9–$2.1 billion (nominal dollars) in 1984–1986 and $1.3–$1.5 billion in recent years (HHS 1995, H-14). The 104th Congress has attempted to greatly cut or eliminate the program, reducing funds to roughly $900 million for 1996; as of this writing, the future of LIHEAP funding is unclear. States are given the option to allot up to 15% of their LIHEAP block grant funds to weatherization efforts, which they typically do (Schlegel et al. 1990). This level may be increased to 25% with HHS authorization. Although there are no restrictions per se on the use of LIHEAP funds for weatherization (aside from income eligibility), many states apply DOE WAP regulations. LIHEAP funding for weatherization, which peaked in 1987 and has since declined (Brown et al. 1993), accounts for about a third of weatherization program funds. LIHEAP funds may be used to exceed the limit of $1,600 per household or to fund certain measures that are not allowable under DOE rules.

From 1978 through 1985, when DOE and LIHEAP were the only two significant sources of weatherization program funds, overall funding was about $250 million, about two-thirds of which came from DOE. Between 1986 and 1988, Petroleum Violation Escrow Account (PVEA) funds, or oil overcharge funds, became a significant source of weatherization funding. In 1988, oil overcharge funds contributed approximately $200 million to weatherization, with total weatherization program funding peaking at $500 million that year. Since 1988, however, the oil overcharge portion of the WAP funds has declined every year, to an estimated $50 million in 1992 (Brown et al. 1993). Overall WAP funding has decreased correspondingly since the late 1980s.

Between 1978 and 1989, about 3.9 million low-income homes were partially or fully weatherized (Brown and Beschen 1992), but approximately 17 million eligible households remain to be addressed. Of the homes weatherized in 1989, 21% were units in apartment buildings (Figure 4-1). Of these, 9.2% were units in large apartment buildings (five or more units), and 11.8% were in small apartment buildings

Figure 4-1

Dwellings Treated Under the
Weatherization Assistance Program (WAP) in 1989

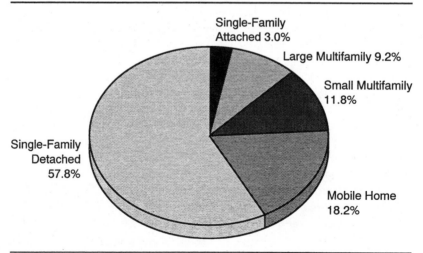

Source: Statistics from Brown et al. 1993.

(two to four units). Of the weatherized large apartment buildings, almost all were centrally heated by gas, electricity, or oil, and about half were located in the New York City area (Brown et al. 1993).

Penetration of weatherization efforts into the multifamily sector has been disproportionately weak. Although 15% of eligible units are located in apartment buildings of five or more units, only 9% of the units weatherized by local WAP agencies are in such buildings. Further, more than half the apartments weatherized in 1989 were located in buildings that were only partially weatherized—that is, buildings in which not all the units were treated (MacDonald 1993).

There are several reasons for the relatively low number of apartment units being treated under WAP. First, at least 66% of a building's tenants must be income-eligible for it to be completely weatherized under DOE rules. The qualification process is resource- and time-intensive, and the rule "causes difficulty with qualifying buildings for eligibility in 6 of 33 states" that responded to a national survey (MacDonald 1993, 11). When a whole building cannot be qualified for treatment, the types of measures that are suitable for installation at the unit level are much more limited.

Another barrier to increasing WAP investment in the multifamily sector is the DOE WAP requirement that landlord and tenants reach agreement as to the length of time before the rent may be raised. This clause, which is in effect a restriction on rent hikes, is intended to ensure that landlords do not reap all the financial benefits from the government-funded improvements. However, this restriction can act as a disincentive to building owners to participate in weatherization programs and is partially responsible for the disproportionately low number of apartment units being weatherized (DOE 1985).

Under WAP, the categories of measures installed in apartment buildings are essentially the same as those installed in single-family dwellings: air leakage reduction; attic insulation; wall insulation; retrofit or replacement of water heating, space heating, and ventilation equipment; windows and doors; structural repairs; and safety and health measures (see Table 4-1). Despite the wide range of measures installed, 80% of materials costs spent on apartment buildings during program year 1989 were spent on windows (MacDonald 1993).

The geographic distribution of apartment buildings treated under WAP is most highly concentrated in the Northeast and the upper Mid-Atlantic region, with moderate activity in the Midwest and certain pockets in the West. Almost no apartment buildings were weatherized under WAP in the South (MacDonald 1993). The state with the highest level of apartment building WAP activity is New York, followed by Illinois and Minnesota. Other states with high levels of WAP activity in apartment

Table 4-1

Measures Installed in Multifamily Buildings Under the Weatherization Assistance Program (WAP) in 1989

Measure	% of Buildings	% of Units
Caulking/weatherstripping	90%	62%
Structural repairs	55%	33%
Door replacement	35%	21%
Other health/safety measures	33%	24%
Low-flow faucet aerators or showerheads	27%	43%
Water heater pipe insulation	26%	21%
Storm windows	26%	11%
Thermal windows	24%	48%
Air sealing	23%	18%
Attic insulation	23%	25%
Water heater tank insulation	17%	11%
Thermostat or other controls retrofitted	17%	13%
Space heating equipment tune-up	9%	27%
Space cooling equipment tune-up	6%	5%
Wall insulation	5%	5%
Storm doors	4%	2%
Ventilation system retrofit/repair	3%	4%
Distribution system retrofit	3%	19%

Source: Adapted from MacDonald 1993, 20 (Table 2).

buildings include California, Washington, Colorado, Massachusetts, Maine, Connecticut, Maryland, and New Jersey (MacDonald 1993).

In New York City and the surrounding counties, where over half of the weatherized apartment units have been located, local WAP providers perform audits and are overseen by the NYC Weatherization Coalition, which performs audits as well as provides quality control. The installations are tailored to each building and often involve mechanical system improvements—about a fourth of the 200 buildings in 1993 were treated with boiler replacements or major boiler system upgrades (Padian 1994b). Per an administrative policy in New York, landlords must contribute at least 25% of the cost of the weatherization service (unless they are income-eligible for WAP services). Several other states also have policies regarding owner investment (MacDonald 1993). In New York, the WAP agencies typically ask the building owner to match the WAP investment on a dollar-for-dollar basis. As a result, many building owners contribute more than the minimum 25%. Despite the owner contribution stipulation, WAP is relatively popular among building owners in New York City, as is evidenced by the fact that, in most cases, it is the building owner who approaches the WAP agency (Padian 1994b). When an owner needs financing, Conserve, Inc., a local nonprofit specialized energy service company, offers a loan packaging service, which is discussed further in Chapter 5.

Little evaluation has been made of apartment buildings treated under WAP, and as a result, programwide data on program savings or cost-effectiveness in apartment buildings do not exist. The only full-program evaluation results available to date are for buildings in New York City and Seattle (MacDonald 1993). The evaluation of New York City buildings involved 570 dwelling units in 12 large oil-heated buildings. The results, unadjusted for any control group, showed an average annual savings of 17%, with a benefit/cost ratio of 2.56 for the DOE investment, not including owner investment (Synertech 1990). Details on the measures installed in these buildings are not available. Okumo (1991) reported extensive evaluation results for all-electric apartment buildings treated under WAP by Seattle City Light. Installed measures typically included windows, venting, caulking and weatherstripping, safety measures, and in some cases, lighting and low-flow showerheads. Net savings were 4–9% of pre-retrofit electricity use (after adjustment for a control group), but the program did not meet the utility's criteria for cost-effectiveness, largely because of high installation and administration costs for the low-income buildings portion of the program (Okumo 1991).

Housing and Energy Conservation Programs

The U.S. Department of Housing and Urban Development (HUD) has also contributed to the improvement of energy efficiency in apartment buildings, both publicly and privately owned. A number of HUD programs providing grant funds to states and communities for community development have adopted weatherization assistance as an eligible expenditure for the grant funds. The HUD grant programs for public housing authority (PHA) assistance, particularly those aimed at modernization, have provided PHAs with weatherization assistance for public housing. The Community Development Block Grant (CDBG) program and the Urban Development Action Grant program (UDAG) have provided weatherization funding for privately owned housing.

The CDBG program was enacted in 1974 to provide communities and states with grant funds for neighborhood revitalization, economic development, or improved community facilities and services. About a third of CDBG funds in larger cities and counties are targeted for property rehabilitation, roughly 20% of which is spent for energy efficiency improvements (Groberg 1994). The CDBG program has also supported, through its technical assistance funding, a 12-year effort to help localities determine feasibility of and preliminary design for district heating and cooling systems. Special attention has been given to connecting systems with public housing (Groberg 1994).

Enacted in 1977 and phased out in 1993, the HUD UDAG program was designed to provide funds to encourage the development of commercial, neighborhood, or industrial projects in cities that were threatened by economic decline. The UDAG funds were granted to communities on the basis of a competitive application process, rather than on a formula of need, as in the CDBG program. The UDAG program gave special attention to energy projects: 34 projects involved low-interest loans for energy improvements, and half of these projects involved residential conservation measures (Groberg 1994).

Although the overall programs are not specifically targeted for energy conservation, a portion of the various funds that HUD allocates to PHAs have been used for weatherization and other energy conservation efforts in public housing. HUD provides funds for several purposes, including grants for constructing new units or acquiring and rehabilitating existing ones and modernization funds for correcting physical deficiencies. Although several programs could potentially contribute to the improvement of energy efficiency in public housing, the two most widely used programs are the Public Housing Modernization Comprehensive Grant Program and the Public Housing Modernization Comprehensive Improvement Assistance Program (CIAP). Both grant funds to PHAs for financing capital improvements in public housing developments, the former program being for developments with 250 or more units and the latter (CIAP) for those with fewer than 250 units. Between 1982 and 1986, HUD provided $756 million for energy efficiency improvements to public and Indian housing authorities under CIAP (ORNL 1992). Most of these funds were used for window replacements and mechanical systems retrofits. Subsequently, there has been a drop in spending for energy efficiency improvements disproportionate to the overall drop that has occurred for modernization funding. Only about 7% of all CIAP funds used from 1989 to 1991 were expended for energy-related capital improvements, which is about half the percentage used for energy measures in the early 1980s (Ashmore 1994). This decline is also related to a gradual shift in emphasis from using modernization funds for energy improvements to employing energy performance contracting, a change catalyzed by a 1987 amendment to the Performance Funding System; see Chapter 5 for a more detailed account of this amendment.

HUD has recently awarded a grant to the National Center for Appropriate Technology (NCAT) to establish a national clearinghouse for technical information and assistance to promote energy efficiency in public and assisted housing. The program will provide training and technical assistance in the areas of energy efficiency and resource conservation to public housing authorities and owners of HUD-assisted buildings. The NCAT program aims to link the existing public housing network with the best technical expertise on improving energy efficiency in multifamily buildings (Masker 1995). The program began operating in 1995 and planned to offer training sessions to public housing officials starting in 1996 (Hayes 1995; refer to NCAT's listing in Appendix B).

In the mid-1980s, in recognition that federal WAP, LIHEAP, and HUD grants could address only a small portion of the vast group of eligible households (and had addressed a small proportion of apartment buildings), interest grew within DOE regarding alternative ways to finance energy conservation activities. A 1986 survey of WAP agencies showed that only 25% of the WAP community action agencies had ever used such innovative financing methods as direct loans, loan interest reductions, leases, lease-purchases, shared savings, energy service contracts, guaranteed cash flow contracts, or other conservation incentives (Lambert 1986). In fact, 40% of the agencies had never heard of such financing methods.

Also in 1986, DOE began to encourage the state and local use of partnerships, resource leveraging, and self-generated resources for low-income weatherization. Through a pilot program entitled Partners in Low-Income Residential Retrofit (PILIRR), DOE granted funds of $100,000 each to five states to develop and implement programs that initiated public-private partnerships for low-income weatherization. The state of Washington initiated an apartment building retrofit program that offered matching funds to entice owners to contribute to the cost of retrofitting their low-income apartment buildings. A study of these programs found that "innovative local groups, with appropriate incentives, can [to a considerable extent] accomplish leveraging through creative packaging of public and utility funds, and use these resources to entice the private sector to participate to a greater extent than in the past" in weatherization initiatives (Callaway and Lee 1988, 37). Such encouragement by DOE for the use of alternative financing mechanisms seems to be a promising initiative that could greatly expand the reach of energy conservation initiatives. These mechanisms are discussed more fully in Chapter 5.

Expired or Canceled Federal Programs or Initiatives

In addition to the above energy efficiency programs, there have been a number of federal programs that have, since their initial implementation, either expired or been canceled. The Residential Conservation Service (RCS), for instance, was a residential energy audit program required by the 1978 National Energy and Conservation Policy Act, and the Commercial and Apartment Conservation Service (CACS) was an audit program established by the Energy Security Act of 1980. The Solar Energy and Energy Conservation Bank, which was also established by the Energy Security Act of 1980, was an entity administered by HUD for solar energy and energy conservation grants and loan subsidies.

The RCS program required large electric and natural gas utilities to offer residential energy conservation audits to their customers. Under this program, utilities provided their customers, upon request, an on-site professional audit. The RCS program, however, addressed neither the issue of funding the installation of the suggested measures nor the availability of the measures themselves (OTA 1992), and the absence of such elements reduced the incentive for customer participation. This situation was the case particularly for residents in apartment buildings, who had little incentive to pay for improvements to rented units. Only a few of the states made an effort to market this program to the multifamily sector (ibid.). Furthermore, the incentive for the utilities was practically nonexistent. Although the utilities covered the entire cost for the audits, as well as the administrative costs for the implementation of the RCS program, state utility regulations generally prevented the utilities from profiting from any resultant energy savings. The cumulative result of these various disincentives was a low customer participation rate for the RCS program. By the time the program ended in 1989, pursuant to its legislative expiration date, only 11% of the eligible population had opted to participate (ibid.). As documented by DOE surveys of the RCS, the participation rate was even lower for low-income and rented (apartment building) households (ibid.).

Although the achievements of the RCS were not as great as had been hoped, the program prompted utilities to become familiar with energy conservation potential, retrofit measures, and the implementation of an energy audit program, and today many utilities continue to provide the energy audit services, either on their own initiative or under direction from the state.

Similar to the RCS, the CACS required large electric and natural gas utilities to offer energy audits for commercial and apartment buildings. Apartment buildings covered under this program included centrally heated or cooled buildings with more than five units. However, the CACS suffered because it lacked the requisite participation incentives. By 1985, only a few states had submitted implementation plans, and only Michigan had actually initiated a CACS program. The following year, Congress repealed the program because of the lack of state interest and the consistently low program fund appropriations (OTA 1992).

After the demise of the CACS, some states modeled their own mandatory audit programs on the CACS program concept. As a result, some utilities had programs similar to the CACS program even after the federal program was canceled. Minnesota, for example, created the Maxi Audit and multifamily training program for use in its PSC-mandated utility programs.

The Solar Energy and Energy Conservation Bank (SEECB), established under the same act as the CACS auditing program and administered by HUD, provided grants and loan subsidies for the purchase and installation of energy conservation measures and solar energy retrofits. Eligible recipients of these funds included apartment buildings, single-family dwellings, and nonprofit commercial and agricultural buildings with low- and moderate-income owners and tenants. The program operated by means of cooperative agreements with states, which disbursed SEECB funds. Of the 55 cooperative agreements (48 states, 5 territories, a group of Indian tribes, and the District of Columbia), 29 programs specifically addressed conservation in the multifamily sector (HUD 1986). About $5.8 million was spent on about 5,300 conservation projects in the multifamily sector, out of a total of $65.9 million spent in the program between 1981 and 1986 (ibid.). The program expired in March 1988, and although it was reauthorized in a different form in 1992, it was never funded (Groberg 1994).

Federal Energy Efficiency Standards

Since the 1950s, the federal government has issued Minimum Property Standards (MPS) for homes and developments using federally financed mortgages, and among these standards are criteria for energy usage and efficiency. Then in the 1970s, as a response to the OPEC oil embargo in 1973, the U.S. Congress initiated various programs to secure domestic energy resources by reducing energy use in all building sectors. Such programs included the development of energy efficiency standards for both buildings and appliances as well as research and development to improve appliance efficiency. Although these programs do not directly target apartment buildings, efficiency standards help save energy as newer, more efficient equipment replaces existing equipment.

Under the Energy Policy Act of 1992 (EPACT), DOE now requires that all new construction of public or federally assisted housing, both single-family and apartment buildings, meet or exceed the Council of American Building Officials Model Energy Code (CABO-MEC) 1992. Public or publicly assisted apartment high rises must meet the American Society of Heating, Refrigerating and Air-Conditioning Engineers (ASHRAE) 90.1 standards. In addition, EPACT requires all states to consider adoption of CABO Model Energy Code for all residential new construction.

Efficiency standards for appliances and equipment in buildings were anticipated by the Energy Policy and Conservation Act of 1975, which required the Federal Energy Administration to develop voluntary targets for appliance efficiency. In 1987, the National Appliance Energy Conservation Act (NAECA) was signed into law, establishing minimum efficiency or maximum energy use standards for certain appliances. Most of the appliance categories are governed by numerical standards, although some are governed by design standards (OTA 1992). Included under the NAECA energy efficiency standards are refrigerators, freezers, room air conditioners, central air conditioners and heat pumps, water heaters, furnaces, dishwashers, clothes washers, clothes dryers, direct heating equipment, kitchen ranges and ovens, pool heaters, television sets, and fluorescent lamp ballasts (ibid.). DOE is required to update the NAECA standards to reflect feasible technical improvements.

EPACT expands the coverage of the appliance energy efficiency standards under the Energy Policy and Conservation Act. The act directs the National Fenestration Council to develop a voluntary rating program for windows and calls for DOE to provide financial assistance and support for a voluntary national testing and information program for luminaries. The act also requires DOE, in conjunction with EPA, to report to Congress on the potential for the development and commercialization of more efficient appliances than currently required under federal or state law.

Federal R&D Programs

As noted in the introduction of this chapter, the first significant government research that specifically addressed energy usage in the multifamily sector was a 1979 study commissioned by the DOE Buildings and Community Services Office in compliance with the National Energy Conservation and Policy Act (DOE 1979). The primary purpose of the report, which was in essence a compilation of data, primarily from census housing figures, was to determine the investment criteria of apartment building owners and to provide owners with information on cost-effective energy efficiency measures. In 1984, DOE completed a draft plan for addressing the multifamily sector; the plan provided information on energy consumption, conservation efforts to date, and barriers to energy efficiency improvements in apartment buildings (DOE 1984).

That same year, DOE consolidated its research retrofit activities into a new R&D program, the Retrofit Research Program, to foster retrofit savings in apartment, single-family, and commercial buildings. Lawrence Berkeley Laboratory (LBL) was named primary lab for the multifamily sector. Following the publication of a multiyear research plan (Diamond et al. 1985), work began on audits, diagnostics, field demonstrations, and evaluations of retrofit technologies in a variety of multifamily building types. LBL analyzed air leakage, ventilation, and energy consumption in a high-rise, all-electric apartment building and in a multistory student dormitory in California (Lipschutz et al. 1983; Feustel et al. 1985). Working with local groups, the LBL team monitored boiler performance and air leakage characteristics in apartment buildings in Chicago and Minneapolis/St. Paul (Modera et al. 1985; Diamond et al. 1986).

Part of the DOE-supported multifamily building research at LBL focused on energy use in public housing, examining financing, baseline energy and retrofit evaluation, and the persistence of savings (Ritschard and Dickey 1984; Goldman and Ritschard 1986; Mills et al. 1986, 1987; Greely et al. 1986, 1987; Vine et al. 1989; Ritschard and McAllister 1992). DOE also supported the development of monitoring protocols to allow different groups to collect data on multifamily (and other) buildings in a compatible format, thus facilitating the exchange of information. The multifamily monitoring protocol developed by LBL was adopted by the American Society for Testing and Materials (ASTM) in 1991 (Szydlowski and Diamond 1989). To promote the use of monitoring protocols, DOE awarded ten grants in a competitive solicitation for retrofit demonstrations, two of which addressed apartment buildings: a study of 350 low-income units in Burlington, Vermont (Diamond et al. 1992), and a study of the persistence of savings from boiler and window retrofits in apartment buildings in New York City (Saxonis 1993).

The Retrofit Research Program changed its name in 1992 to the

Existing Building Efficiency Research Program (EBER), and DOE continues to support work in the multifamily sector. Much of this research is made possible through the DOE-HUD Initiative, which addresses the entirety of the federally assisted housing stock. Between 1990 and 1994, the initiative funded 26 projects to improve energy efficiency in federally assisted housing. A key goal of this work was to expand public-private linkages to improve energy efficiency in this sector. An expanded version of the initiative was planned to begin in 1996, but as of this writing, these efforts are jeopardized because of shifts in priorities by the 104th Congress.

State Programs and Initiatives

As with federal programs, few state energy efficiency programs specifically target the multifamily sector. State-level mechanisms that may address energy efficiency in apartment buildings include efficiency standards, informational programs, targeted audit and installation programs, financial incentives, and research funding.

Efficiency Standards

The implementation of energy efficiency standards has proven to be problematic in the multifamily sector and frequently encounters political opposition from various interests. Not only do some building owners object because such standards could require them to make costly improvements to their properties, but also some advocates of low-income housing object to some regulatory requirements for fear of causing low-income housing rent costs to rise above affordable levels.

The Minnesota experience with implementing mandatory rental unit energy efficiency standards illuminates some of these obstacles to the implementation of regulatory standards (Altman 1981; Hubinger 1984). The Minnesota Energy Conservation Standards for existing residences, enacted in 1976, called for a mandatory program for rental units called the Residential Rental Retrofit program (RRR). The RRR required that year-round rental residences be brought into compliance with specified prescriptive energy conservation standards, provided that they did not result in paybacks of greater than ten years. The required standards included weatherstripping on exterior doors and windows; caulking or sealing of exterior joints and openings in the exterior envelope; installation of positive shut-offs for all fireplaces and fireplace stoves; insulation of accessible attics (R-19), accessible rim joist areas (R-1), and accessible walls and floors (R-11); and the installation of storm doors and windows (Hubinger 1984).

One of the primary drawbacks of the RRR was that the prescriptive envelope standards were uniform for both small and large housing developments, and many heating, cooling, and ventilation system measures were ignored. Due to their inappropriateness for apartment buildings, these standards were considered weak and ineffective. In addition, enforcement of the law presented several problems. The legislature had failed not only to appropriate funding to inform owners of the new requirements but also to appropriate funding to the Department of Energy and Economic Development for compliance enforcement. City inspectors did not have authorization to inspect for compliance unless the individual city councils adopted the standards into their own codes. Finally, the absence of direct financial incentives in the RRR program, in conjunction with poor enforcement by the appropriate agencies, gave rental property owners little incentive to comply with the law.

Informational Programs

A look at the energy efficiency program history in the state of Wisconsin provides insight into the critical features of an informational certification program. In 1980, after the Wisconsin Public Service Commission (PSC) considered establishing a mandatory residential conservation program and met formidable opposition, the PSC opted instead to order a voluntary residential conservation program entitled the Voluntary Rental Living Unit Conservation Program (RLU). RLU program elements were delineated in a general form by the PSC, and the ten largest state utilities were obligated to implement them. Each utility was required to offer rental housing customers a variety of services: a PSC-approved energy conservation standard, a structural audit of the living unit, a "lifestyle" energy audit of the tenants' behavioral characteristics, and a seal and certificate program for those units that met energy efficiency standards (Fay 1984). In addition, each utility program was required to have an advisory committee and outreach group that assisted in the planning and implementation of the program, as well as extensive media advertising to educate consumers about the availability of the RLU program (ibid.). The focus of the RLU program was to advertise the seal and certificate as a useful measure of energy efficiency for both tenants and owners so that tenants would be persuaded to look for it and owners would be persuaded to try to qualify for it.

Buildings with five or more living units were given two criteria options for meeting the standards required for the receipt of the energy efficiency certificate. The first option consisted of meeting prescriptive standards as required of small apartment buildings: ceiling insulation between R-19 and R-38, insulation over attic access hatches, caulking, weatherstripping, storm windows and doors, box sill insulation of R-19, insulation of all crawlspace areas, and ventilation for attic areas (Fay 1984). The second option, generally preferable for large apartment buildings, consisted of meeting a performance-based standard. In this, the RLU program was responsive to the different energy performance characteristics of the multifamily sector and acknowledged that large apartment buildings might benefit from energy efficiency measures that are different from those employed by the sector of smaller apartment buildings. The RLU recognized that larger apartment buildings have a greater potential for achieving energy efficiency through improvements applied to the mechanical systems, rather than improvements to just the building envelope.

The performance standard required that design heat loss, excluding infiltration and ventilation, through walls and ceilings not exceed 13 Btu/hr/ft^2 for the total building envelope. According to Fay (1984), the efficiency measures the Wisconsin Gas Company most commonly used in achieving the performance standard include:

- Improving boiler efficiency through periodic cleaning and tuning
- Cleaning outside combustion air louvers to ensure unobstructed intake
- Recalibrating all control systems
- Developing a maintenance plan for steam traps
- Repairing leaks in domestic hot-water and hydronic heating piping
- Increasing attic insulation to R-38
- Caulking and weatherstripping windows and doors
- Providing storm doors and windows
- Installing air conditioner covers during the heating seasons

A critical feature of the RLU program was the extensive marketing campaign that accompanied the implementation of this program. Wisconsin Gas employed a direct-marketing approach that targeted landlords within its service area, achieving a response rate of 25% (Fay 1984). This utility also used a television and radio marketing campaign, which it found to be effective in reaching tenants but not landlords.

In 1985, the voluntary RLU was suspended when Wisconsin's Department of Industry, Labor and Human Relations implemented a program of mandatory Rental Unit Efficiency Standards. The code delineating the mandatory standards prescribed certain conservation measures that must be in place before rental property can transfer ownership. One drawback of this program is its reliance on prescriptive measures; this approach ignores the fact that the best conservation measures are different for each building. This distinction may be critical for large apartment buildings that could benefit from heating or cooling system retrofits. The Rental Unit Efficiency Standards program has also been criticized for its accessibility exclusions. For example, if the installation of insulation requires that a hole be drilled in a wall or a hatch created for access to the attic, the insulation is deemed unnecessary. Finally, resources for enforcement have been insufficient, and compliance has been problematic (Berkowitz 1994).

When the RLU was suspended and the Rental Unit Efficiency Standards program began, the Wisconsin PSC felt it was important that utilities continue to provide information to allow comparison of the energy efficiency of apartments (Berkowitz and Newman 1988). To meet this need, the PSC designed the voluntary Wisconsin Heating Energy Efficiency Label (WHEEL) rating system. The WHEEL is an energy rating system that uses a relative scale to rate building efficiency according to building heating energy consumption in terms of $Btu/(ft^2 \cdot DD)$ ($Btu/ft^2/degree$ day). Different methods of calculation are used for small and large apartment buildings in recognition of their different energy use characteristics and potential. The rating label includes information concerning how the building compares with an average structure and a chart reflecting approximate space heating costs for an average unit in the building. In the past few years, however, utility funding of the program has been very limited (Berkowitz 1994).

State Research Programs

Funded at $15.5 million a year, the New York State Energy Research and Development Authority (NYSERDA) is the largest state R&D program in the United States (Harris et al. 1992) and has been investigating the intricacies of energy use in apartment buildings for over a decade. In the 1980s, NYSERDA developed an audit, worked toward identifying institutional barriers to energy efficiency improvements in apartment buildings, determined the cost of implementing an energy conservation program for publicly assisted housing in New York State, and conducted technical studies of electrical submetering systems in high-rise apartment buildings and of the energy- and water-saving potential of low-flow showerheads (DOE 1985). In addition, NYSERDA undertook an effort to demonstrate energy savings that can be achieved in apartment buildings through performance contracting (ibid.).

In the 1990s, NYSERDA has continued its research efforts to examine energy use in apartment buildings. It has implemented at least 14 research projects related to energy use in the multifamily sector since 1990, some of which are still ongoing. Areas of study include domestic hot-water consumption and efficiency analysis, domestic hot-water system sizing, air flows, steam-to-hydronic conversions, air leakage, thermostatic radiator valves, holistic energy audits, heating control systems, the applicability of the Princeton Scorekeeping Method (PRISM) to apartment buildings, decentralized space and water heating, the use of blower doors in apartment buildings, ventilation, information transfer, and improving the energy efficiency of public and publicly assisted housing. The total investment for these 14 apartment building research programs is expected to exceed $2.28 million (Karins 1994).

Other State-Sponsored Programs

In 1988, the Energy Resources Center in Chicago created the Energy Efficiency Rehabilitation program, under which abandoned or near-abandoned gas-heated low-rise brick apartment buildings are "superinsulated" during a larger substantial rehabilitation process. The Illinois Department of Energy and Natural Resources provides the funding, and the Energy Resources Center provides technical assistance, with several nonprofit organizations playing various roles in the project. The superinsulation process costs roughly $2,000 per unit and involves the installation of R-43 attic insulation and other techniques, plus high-efficiency heating systems (see the Energy-Efficient Building Rehabilitation case study in Chapter 3). By the end of 1993, 100 units in 7 buildings had been superinsulated, resulting in the gas consumption in at least half those units being cut by 80–90% (Knight 1993–1995). As of December 1994, about 220 units in 20 buildings had been treated (ibid.). In the future, the program will continue to focus on substantial rehabilitation but will begin to include some moderate rehabilitation as well, which will emphasize air sealing (ibid.).

Since 1986, the Vermont Housing Finance Agency (VHFA), in partnership with the nonprofit Vermont Energy Investment Corporation, private developers, and government programs, has implemented energy efficiency programs for VHFA-financed housing. One such program is the Multi-Unit Rental Housing Energy Efficiency Program. In 1987 and 1990, VHFA identified housing developments with the highest per-unit energy costs from among 100 multi-unit housing developments for which VHFA is a mortgage holder. VHFA informed the owners about the savings opportunities possible from energy efficiency improvements and identified energy service companies that could provide a comprehensive range of services for the implementation of the energy efficiency program. The scope of work of these energy service companies covers building energy analysis, cost estimates, cash-flow analysis, construction management, preparation of work specifications, bid preparation, contractor selection and oversight, and final inspection. Some of the energy efficiency measures installed in multi-unit VHFA-financed housing include high-efficiency boiler and water heater retrofits, apartment thermostat upgrades, insulation upgrades, air sealing, and water conservation measures. In some cases, fuel switching from electric storage heat to natural gas has been prescribed. These measures have resulted in significant energy savings for several VHFA developments. The innovative financing arrangements involved with this program are discussed in Chapter 5.

Local and Regional Programs

A number of energy efficiency initiatives serving apartment buildings have been implemented by municipalities and by locally based nongovernmental organizations. Although cities and towns can be limited in their ability to implement programs in energy conservation because of constrained financial resources, they are also very promising entities for the implementation of such programs because of their close contact with local housing issues and needs and their knowledge of the local housing stock characteristics. Nongovernmental organizations, working closely with or spawned by city programs, have proven to be a very creative force in energy conservation for apartment buildings as well as for single-family housing.

Municipal Energy Efficiency Programs

Several local governments have adopted codes and standards that include attention to energy efficiency in buildings. For example, in the mid-1980s, the city of Madison, Wisconsin, had mandatory energy efficiency standards for new construction. And in San Francisco and Berkeley, California, residential buildings must meet a specified energy efficiency standard before they may be sold to another owner. In Portland (Oregon), Minneapolis, and Madison, owners of apartment buildings must install a series of prescriptive measures before they sell the building. However, compliance with most of these programs has been problematic (DOE 1985). Such compliance problems tend to result from inadequate enforcement and lack of incentives. The city of Chicago took an innovative approach to increasing the energy efficiency of apartment buildings when it passed an ordinance in 1988 requiring landlords to give prospective tenants an estimate of the apartment's heating costs, based on a 12-month history. The intended effect of the ordinance was to induce building owners to make their apartments more competitive by reducing their energy costs.

In 1983, the city of Tacoma, Washington, became the first local government in the Pacific Northwest to adopt the Model Conservation Standards (MCS) established by the Northwest Power Planning Council as part of the Northwest Power Plan (Bellamy and Fey 1988). The MCS is a building energy performance standard for new residential and commercial buildings utilizing electricity for space conditioning. To implement the MCS, the city of Tacoma took four major steps. First, it created an information center that, in alliance with the local Energy Extension Service Resource Center, disseminated information and answered questions about MCS and energy-efficient construction. Second, it developed a program of training sessions on energy-efficient construction techniques and a manual for guidance in this area. Third, it formed a financial assistance/incentive program, which grants a sum of money to a building owner upon the successful completion of the inspection process establishing the attainment of the MCS standards. Finally, the city launched a marketing campaign to create a strong market for energy-efficient homes. By 1987, nearly 6,000 apartment units in dwellings ranging from triplexes to 600-unit complexes had been built using the standards or were under plan for review (ibid.).

In 1987, the Bonneville Power Administration produced the Northwest Energy Code (NWEC), based on the MCS. Four years later, similar standards became mandatory statewide as the Washington State Energy Code. Local governments that had adopted NWEC before the state energy code was enacted may continue to enforce NWEC for electrically heated buildings and homes.

In 1992, Tacoma City Light evaluated the MCS program's compliance levels and energy use impacts through field audits, review of administrative records, and interviews with builders and code officials (Perich-Anderson and Dethman 1994). The study found a compliance level of 91% for residential buildings. Another study which examined the persistence of savings in apartment buildings built according to MCS found first-year savings to be 7–15% of reference building consumption, with savings persisting through the second year of the study (Schuldt et al. 1994). However, these figures exclude the effect of air-to-air heat exchangers, which had a large negative effect on energy savings.

Center for Energy and Environment, Minneapolis

Beginning in 1981, a group of energy conservation practitioners, engineers, and researchers in Minneapolis, Minnesota, pursued a buildings energy efficiency program that evolved into one of the country's exemplary efforts in retrofit of apartment buildings. Originally started with a HUD Innovative Grant plus support from the local gas utility, Minnegasco, the programs operated as part of the Minneapolis Energy Office (MEO) but in 1989 became a separate nonprofit entity now known as the Center for Energy and Environment (CEE). One notable aspect of this program has been its emphasis on field testing of retrofit strategies and careful evaluations based on the actual performance of the recommended retrofits. Targeting is one of the hallmarks of an effective program, and the CEE work is region-specific, focusing largely on two- to seven-story walkup apartment buildings, which dominate the multifamily stock in the Minnesota Twin Cities area.

Most apartment buildings covered by CEE were less than 60 units in size and were privately owned. CEE did some work in larger apartment buildings and audited Minneapolis public housing but were able to implement few retrofits in the public housing because of lack of interest by the housing authority. However, CEE did succeed in reaching a substantial portion of the privately owned apartment buildings in Minneapolis and surrounding areas. When carefully interpreted, much of the MEO/CEE technical work (cited throughout Chapter 3; in particular, see Hewett et al. 1994) generalizes to similar buildings and equipment in other locations. MEO/CEE collaborated or coordinated with others involved in conservation, such as the Energy and Environment Research Center in St. Paul, the Center for Neighborhood Technology in Chicago, other leading private energy conservation specialists, local utilities, and researchers at universities and Lawrence Berkeley Laboratory.

The approach pioneered by CEE for implementing energy efficiency improvements illustrates the services that can be provided by municipal or nonprofit programs to effectively reach the multifamily sector. CEE has assisted with gas and electric utilities in carrying out demand-side management (DSM) efforts and has developed a particularly good model for serving private owners of apartment buildings. CEE furnishes a "one-stop shopping" service for building owners to provide building audits and retrofit planning, helping identify cost-effective retrofits, specify equipment, select contractors, oversee installation, and assure quality control for the building owner. In addition to careful retrofit testing and pilot programs used to develop a reliable information base on effective retrofits, CEE's program

143

worked with equipment suppliers and mechanical contractors ("trade allies") to provide engineering assistance and training as well as to introduce and promote new technologies. The result is a pool of qualified contractors who provide prescreened equipment specifications and prices, which are most important in providing building owners with assurance that the job will be done right and that their money will be well spent. CEE's early projects included inspection and follow-up work on the contractor installations. Cost and resource constraints inhibited such quality control in all of the work; however, CEE still considers such "retrofit commissioning" to be a valuable aspect of technical assistance. State and municipal housing finance agencies provided packages of low-interest loans for energy efficiency upgrades, which CEE helped market to building owners. CEE found that such financial incentives, plus rebates from utilities, are crucial for addressing the up-front cost barriers faced by a building owner considering retrofits.

The MEO/CEE program has audited over 2,700 buildings since 1981, covering roughly 70% of the multifamily housing stock in Minneapolis proper and about 30% of the stock in adjoining suburbs. Historically, the program realized a 67% completion rate for its recommended measures. Recently, however, this rate has dropped to about 30% as the program has saturated the local market. Savings from the installed measures are substantial, amounting to a cumulative total of over 10 million therms since the beginning of the program. Retrofit expenditures averaged $700 per building, and annual savings averaged approximately 1,000 therms per building, or about 7% of pre-retrofit gas use (Lobenstein 1995). The MEO/CEE program has served as a model for gas utility DSM programs in other states.

Having essentially covered the multifamily stock in metropolitan Minneapolis, CEE's current multifamily programs are oriented toward working with utilities to serve apartment building customers in far suburbs and completely outside the metropolitan area. Unfortunately, funding for such work has become more limited than it previously was. Nevertheless, the MEO/CEE efforts from 1981 through the early 1990s leave a rich legacy of technical and programmatic wisdom. Their expertise remains an important resource for others seeking to pursue energy efficiency programs for multifamily housing.

Environment and Energy Resource Center, St. Paul

Established through a partnership between the city of St. Paul and Northern States Power and originally known as the Energy Resource Center (ERC), the Environment and Energy Resource Center (EERC) is a private, nonprofit corporation also active in energy conservation in Minnesota. EERC has administered energy conservation programs, including shared-savings projects, in about 300 apartment buildings since 1986. In a recent analysis of EERC's work, 23 buildings with multizone heating systems and 20 with single-pipe steam were evaluated with PRISM and showed payback periods of under three years (Bryan 1995).

Center for Neighborhood Technology, Chicago

Throughout the 1980s, the Center for Neighborhood Technology (CNT) in Chicago was a leader in apartment building energy efficiency retrofits as part of its broader community development efforts. Between 1982 and 1984, CNT developed and managed energy efficiency retrofit services for 400 apartment building projects. Then, from 1984 to 1989, CNT operated the Chicago Energy Saver's Fund (CESF), sponsored by Peoples Gas Light and Coke Company (Peoples) and the city of Chicago. Using a one-stop shopping approach, the program assisted owners of apartment buildings (typically three-story walkups) with the process of loan applications, detailed energy audits, contractor bids, and energy use monitoring. With program loan funds, building owners would invest an average of $1,400 per unit on energy conservation measures (Bernstein 1994). The most commonly installed measures were, in declining order, storm windows, indoor thermostats, boiler replacements, radiator work, and ceiling cavity insulation (Graham et al. 1991). According to an evaluation performed by the Wisconsin Energy Conservation Corporation (described in detail in The Peoples Gas Light and Coke Company case study later in this chapter), energy savings from the program averaged 28% (Graham et al. 1991). However, despite the impressive savings, the loan program was discontinued in 1989 after Peoples found the program was not cost-effective from the utility perspective. CNT no longer performs audits or direct installation of retrofits since utility and city funding for conservation ended in 1992.

Energy Office, Portland, Oregon

In 1987, the Portland Energy Office launched its Multifamily Weatherization Program to induce building owners to weatherize apartments. Developed with funding from the Bonneville Power Administration and the Oregon Department of Energy, the program draws on existing state and utility incentives and utility audit capabilities to provide one-on-one assistance to property owners and managers. The Energy Office explains the benefits of weatherization to owners and managers, arranges for energy audits, consults on audit recommendations, discusses financing options, and assists with soliciting contractor bids. Owner cost-shares are leveraged with state and utility rebates, loans, and tax credits to form a package to finance the measures.

Energy efficiency measures for electric- and gas-heated buildings qualify for cash rebates of 25% of the cost-effective job cost, up to $1,250 and $350 per unit, respectively, plus 28.87% as a utility pass-through of the state business energy tax credit. Oil-heated buildings receive a 25% cash rebate, plus a 35% state energy tax credit (O'Keefe 1994). Owners may select the building contractor and choose which measures to install, which typically include a combination of windows (75%), followed by attic insulation (53%), floor insulation (38%), and wall insulation (15%) (ibid.). Other measures include lighting and water heaters, for which local utilities offer rebates, and low-flow showerheads provided by utilities. An evaluation conducted in 1991 showed that the program generated average space heating energy savings of 26%. By fuel type, average savings were 24% for gas-heated buildings, 35% for oil-heated, and 19% for electrically heated buildings (Marsh Technical Services 1991). By mid-1994, the program had resulted in the weatherization of over 10,000 units of Portland's large apartment buildings (O'Keefe 1994).

In the fall of 1990, the Portland Energy Office conducted an investigative pilot project entitled Energy Savings from Operation and Maintenance Training for Apartment Boiler Systems, funded by the Urban Consortium Energy Task Force (UCETF). The UCETF funds proposals that will directly improve local government services or the revenue base of participating cities (Norton and Lindberg 1992). Portland's apartment boiler systems project entailed operations and maintenance (O&M) training of the operators of boiler heating systems in ten low-income apartment complexes. As noted in Chapter 3 (see Boiler-Based Systems, Tune-Up), this pilot project demonstrated an average savings of 10%.

Utility Energy Efficiency Programs

Utilities are appropriate entities for the implementation of energy efficiency programs for several reasons. First, they not only have direct access to energy customers and their fuel consumption information, but they also have the resources and expertise to understand and respond to the variable characteristics of their customer population. In addition, when the mix of financial incentives and high customer participation is provided by the utilities, the implementation of energy

The Margolis Apartments in Chelsea, Massachusetts, were the site of a utility demand-side management demonstration of energy efficiency retrofits in public housing.

efficiency programs can also be in the utility's best financial interest. However, the historical development of utility programs illustrates the fact that utilities have not always recognized these opportunities.

To the extent that utilities had low-income energy efficiency programs during the 1970s, these programs were usually based on a rationale of customer service, the promotion of social equity, and the reduction of arrearages. Another important motivating force behind utility involvement in early weatherization and other energy efficiency programs was the need to meet requirements established by state public utility commissions or other entities at the federal, state, or local level. In the late 1970s and early 1980s, however, there was a growing recognition that energy efficiency programs and services could be viable least-cost alternatives to the purchase of new supply-side resources. In 1980, when Congress enacted a federal statute (PL 96-501) defining the conservation of electricity as a "resource" that could be purchased by utilities in lieu of new electrical generation, the idea that energy conservation could be substituted for new energy supplies was granted official recognition and greater political acceptance (Eckman et al. 1992). Prompted by the uncertainty of future energy demand, power plant siting constraints, and environmental regulations, this new concept of utility resource acquisition and planning led to the development of "least-cost planning" (LCP), or as it is alternatively called, "integrated resource planning" (IRP), in which demand and supply options are evaluated together in order to best meet customer needs at the lowest cost. In some areas, emphasis on LCP has led in turn to the aggressive promotion of DSM and energy efficiency programs.

In addition to DSM being a cost-effective alternative to supply-side options, utilities have several other incentives for investing in energy efficiency. These include reducing arrearages, improving customer relations, and retaining existing customers who might otherwise switch fuels. DSM investments in residential energy efficiency are also a way for utilities to benefit their local communities by making housing more affordable and helping prevent abandonment. Opportunities exist for partnerships between utilities and community-based organizations—for example, by piggybacking DSM services onto existing weatherization services provided through community action agencies (CNT 1992).

The level of DSM expenditures in the multifamily sector has grown since the mid-1980s. There are currently more than 1,000 utility-run programs for energy efficiency in the residential sector (OTA 1992), providing customers with rebates, information, audits, direct installation, or a combination thereof. Since 1986, at least two dozen utilities have designed and implemented DSM programs targeted directly toward the multifamily sector. However, the effectiveness of utility programs for the multifamily sector varies greatly, depending on such factors as program design, the degree to which the program is specifically targeted to this sector, and availability of financial incentives to encourage participation.

Evolution in utilities' approaches to demand-side services is resulting in a focus on more cost-effective programs. Although this situation may mean fewer direct utility investment dollars, DSM spending can be creatively applied and more highly leveraged, ultimately resulting in larger efficiency investments and larger overall energy savings. Examples of the diverse but creative approaches that can be taken include the Public Service Co. of Oklahoma's technical assistance and incentives for efficiency improvements and groundsource heat pumps at an apartment complex in Tulsa; Pacific Gas & Electric's bid for demand reduction from efficient equipment retrofits in Bay Area apartment buildings; and the New York Power Authority's initiative for stimulating commercialization of apartment-scale superefficient refrigerators for New York City public housing. Large customers that they often are, both public and private apartment buildings make good targets for utility-leveraged investments in efficiency upgrades as a way to competitively include conservation resources within a diversified, least-cost energy mix.

Energy Efficiency Rebate Programs

In order to reduce the consumer's cost of procuring energy-efficient appliances and measures, many utilities offer their customers financial incentives in the form of rebates. Many rebate programs are targeted at a certain large sector of the customer population, such as the commercial/industrial sector or the residential sector, depending on the appliance or efficiency measure covered by the rebate. However, rebate programs typically do not target subpopulations of the residential sector.

As rebate programs usually depend on the utility customer's own initiative in responding to the rebate offer by purchasing the rebated item, it is important to address marketing, education, and customer targeting as the means of obtaining customer participation. Marketing for most utility rebate programs consists of inserts in the utility customers' monthly energy bills. Rebate programs are likely to be less attractive to renters because they are rarely given adequate incentive to pay for improvements to an apartment that they do not own. And many appliances, such as refrigerators, washers, and dryers, are most often purchased by building owners. If rebate and appliance programs are to significantly affect the multifamily sector, they must appropriately target residents, building owners, and managers with sufficient incentives to encourage their participation.

The city of Austin's municipal electric utility offers one of the relatively few rebate programs that specifically addresses the multifamily sector. Austin's rebate program offers substantial rebates for ceiling insulation, window treatments, air infiltration control, lighting, and replacement air conditioners and heat pumps in apartment buildings with four or more units. For R-26 ceiling insulation, the program offers $0.10–$0.15/ft^2, depending on the previous R factor. Window rebates are $1.00/ft^2 for windows facing east, southeast, south, southwest, west, or northwest. For air infiltration control, the program pays $0.07/ft^2 of conditioned space. Lighting rebates are $18 and $15 per fixture, respectively, to convert the kitchen and bathroom fixtures from incandescent to fluorescent lighting. For common-area lighting, the rebates are $7 per fixture for hard-wire retrofits, up to $12 to replace an incandescent fixture with a fluorescent fixture. Additional rebates are offered for ballast replacement, optical reflectors, and occupancy sensors. In occupied units, replacement air conditioner units with a seasonal energy efficiency rating (SEER) of at least 11.0 earn from $400 to $530 in rebates, whereas rebates for heat pump replacements with an 11.0 SEER range from $450 to $560. Rebates for replacement air conditioners and heat pumps in vacant units undergoing major rehabilitation are on average about 30% less than rebates for equipment in occupied apartments (Austin 1994). This program has not yet been evaluated.

Utility Audit and Installation Programs

Utilities have experimented with a wide variety of programs that fit loosely under this heading: audit programs with and without incentives, low-income grant weatherization programs, and direct installation programs. To reach apartment building owners and persuade them to participate in programs, many utilities operating audit and installation programs have implemented comprehensive marketing campaigns specifically targeted to apartment building owners. The most effective marketing campaigns recognize the split incentives between tenants and landlords and understand that owners and managers of apartment buildings are primarily interested in increasing revenues, reducing expenses, minimizing the day-to-day problems in managing their building, adding value to their property, and reducing tenant turnover. Accordingly, marketing campaigns targeted to the multifamily sector have emphasized the ability of energy efficiency improvements to increase cash flow and profitability, improve tenant comfort, and reduce maintenance problems. Utilities have carried out this strategy by contacting and educating owners and managers through direct mailings and phone calls, workshops, meetings, and program presentations.

The most basic utility programs are informational and do not offer customers financial incentives for the purchase of the recommended energy efficiency measures and technologies. The rationale behind this approach is the assumption that residents who have specific information about the energy efficiency potential of their household or apartment building will invest in and install recommended measures without a financial incentive from the utility. Unfortunately, the up-front cost of recommended measures is often enough to prevent residential program participants from installing the measures. In light of this fact, many utilities now offer audit participants financial incentives in the form of grants, rebates, or loans. Although there is a dearth of definitive information concerning the effect of offering financial incentives along with the residential energy audit, some evidence has surfaced indicating that participation rates and energy savings do rise for programs that offer financing options in addition to the audit (Nadel 1990). A recent study that examined evaluation results of hundreds of utility DSM programs supported the notion that incentives are necessary to achieve participation goals (Mast and Ignelzi 1994). The authors, however, stressed that although incentives are necessary for successful DSM, they are not sufficient, as other factors—such as risk, aesthetics, convenience, and transaction costs—are also influential in the customer's decision-making process.

In the 1980s, utilities began to offer low-income residential auditing and weatherization grant services on a more widespread level, driven in large part by public utility commissions citing low participation rates in audit programs and low- or zero-interest loan programs. The weatherization services—which utilities often contract out to community action agencies or energy service companies—may include attic insulation, wall insulation, lighting, refrigerators, water heater blankets, high-efficiency furnaces, low-flow showerheads, duct wrap, weatherstripping, caulking, storm windows, and other infiltration reduction measures. Utilities benefit not only from reduced loads but also from considerable arrearage reduction, reduced costs of bill collection, disconnection, reconnection, and public relations recognition. Despite these benefits, a 1991 survey found that over 40% of states still had no utility participation in low-income programs (Fenichel 1992). Furthermore, these types of programs tend to benefit single-family residences disproportionately more than apartment units.

In recent years, some utilities have begun to offer more comprehensive services that include audits, measure installation services, and financing options for participants. The number of specifically targeted multifamily programs in this area is growing, and they have shown promise for achieving significant savings. With this type of program, the utility (or its contractor) is typically involved in all stages of the process, from the initial building assessment audit to the final building inspection. Operations and maintenance training is typically provided to the apartment building manager or maintenance staff, and workshops or other energy education is often offered to the tenants. A final building inspection and post-retrofit monitoring of the energy savings may also be components of such comprehensive audit and installation programs. Financing options, such as grants, rebates, or loans, are typically offered with these programs; a number of utilities have implemented innovative financing arrangements that include the sharing of the energy savings among the utility, the building owner, and the building tenants. To ensure that the owner or landlord does not try to recoup the costs of participating in the program by raising rents, some utilities' energy service agreements stipulate that rents

may not be raised during the duration of the program, nor for a specified period thereafter.

The Multifamily Conservation Program implemented by Seattle
City Light is a particularly successful apartment building energy audit
and installation program. This program, which entered its pilot phase
in 1986, offers financial and technical energy conservation assistance
to owners of electrically heated apartment buildings with five or more
units. The buildings served are generally low-rise with wood-frame or
concrete construction. Seattle City Light put forth a challenging plan
for the implementation of the program: 48 apartment buildings were
to be served in the first year, and twice that number in each year from
1987 through 1992. Thereafter, the annual building quotas would rise
to a total of 180 (Okumo 1990). The final program goal is to treat 67%
of the apartment buildings in the multifamily sector of the Seattle City
Light service area by the end of the year 2004. This effort would con-
sist of serving 1,934 standard-income and 658 low-income apartment
buildings.

Under the program, Seattle City Light performs building energy
audits to determine the appropriate conservation measures, provides financial and technical conservation assistance, and conducts
extensive post-retrofit monitoring of the program for a study aimed
at determining total program savings. Energy efficiency measures
include double-glazed replacement or conversion windows, attic or
roof insulation, under-floor insulation, wall insulation, caulking and
weatherstripping, low-flow showerheads, water heater wraps and
temperature setbacks, pipe and duct wraps, additional cavity venting, and lighting modifications. Owners of buildings in which two-
thirds or more of the tenants have low incomes are offered a grant
for the full cost of the program, provided that the owner does not
raise rents for 5 years. Owners of buildings with standard-income
tenants are given the option of a ten-year, zero-interest loan from
Seattle City Light, with a five-year deferred payment and a 50% discount for full loan payment in the first year. An evaluation of the
Seattle City Light program can be found in the Program Evaluation
section of this chapter.

Niagara Mohawk Power Corporation in New York offers a program that targets electrically heated low-rise buildings with five or
more units, occupied by low- and moderate-income households.
Qualifying buildings must meet a relatively high threshold of electricity use, 15 kWh/ft^2. The program combines some of the features of
direct investment and performance contracting. For example, the utility approves and pays for the installation of all cost-effective measures, which may include hot-water, controls, envelope, and lighting

measures, and the company contractor receives full payment only if 85% of promised savings are delivered. Over a three-year period, the utility will spend $12.5 million serving 6,000–7,000 units, with an expected 5.8 MW reduction in demand.

Some utilities have programs specifically targeted toward public housing. In 1990, Connecticut Light and Power Company (CL&P) conducted a pilot program with the Willimantic and Danbury housing authorities. The pilot program addressed a 100-unit high-rise building for the elderly, owned by the Willimantic Housing Authority, and a 60-unit two-story town house development owned by the Danbury Housing Authority; both developments are all-electric. Under the terms of the program, CL&P provided a grant approximating 20% of the value of the qualifying measures and a zero-interest loan to cover the remaining 80%. CL&P also paid for the costs of the audit of the buildings and a portion of its contractor's costs in assisting the PHA to secure a waiver of the old performance funding system (PFS) energy-related regulation. The total CL&P contribution was approximately 50% of the project cost. Qualifying measures included attic insulation, window replacements, set back thermostats, dampers and controls on a rooftop exhaust system, air sealing, low-flow showerheads, and hot-water pipe insulation. Investments per apartment were approximately $4,000 at Danbury and $1,300 at Willimantic. CL&P's contractor, an energy service company (ESCO), executed a performance contract with each of the PHAs for a 12-year term. The PHAs separately contracted with the ESCO to provide equipment troubleshooting, annual resident education sessions, maintenance staff training, and performance monitoring at the properties. Any net savings achieved by the PHAs are theirs to keep for the duration of the contract with the ESCO. Net savings accrue whenever the sum of the post-retrofit fuel bill and the debt service to CL&P are exceeded by the pre-retrofit consumption level times the current kWh price. HUD agrees to pay the PHA the latter amount every year. The ESCO guaranteed the projected savings, and if the savings fail to meet projections, the ESCO is liable to pay CL&P the remainder due on the debt service payment.

Early in 1992, the Boston Edison Company (BECo) launched a public housing DSM program directed at 1,350 units in 13 developments managed by 12 PHAs in its service territory. All of the buildings are electrically heated; most are high-rise structures for the elderly. The utility paid for audits, installation of all eligible measures (those having lifetime savings that exceed the utility's avoided costs), and remaining soft costs (those other than equipment or labor costs—for, example specifications development and construction oversight). Hot-water measures, air sealing, and lighting measures were eligible for full BECo subsidy in virtually every building addressed by the program.

In the first year of the program, BECo paid up to its avoided costs for storm windows and window replacements that meet a minimum 2.2 R-value requirement. In most cases, the BECo cost share would be slightly greater than one-half the installed cost of the measures. PHAs managing state-owned public housing properties borrowed their cost-share from the state housing agency that financed their construction. The housing authorities will retire the loan from the energy savings generated over the ten-year period of the loan. The BECo contractor independently guaranteed the savings to each PHA.

Electric/Water Utility Conservation Cooperation Programs

Another recent development in the area of energy efficiency includes initiatives in joint conservation programs between electric and water utilities. This recent interest in cooperation between water and electric utilities stems from an increased awareness that reductions in water use can also result in reductions in energy use for heating, treating, and pumping water. This development may be of special interest to apartment building owners because domestic hot water (DHW) in apartment buildings accounts for a relatively large proportion of energy usage, compared with DHW in single-family residences. In addition, the escalating costs of water in many areas is becoming an increasingly large expense for building owners.

Efficiency measures that could provide both energy and water savings include low-flow showerheads, faucet aerators, and water-efficient clothes washers and dishwashers. When water must be pumped several floors up, low-volume toilets can save electricity as well as water. In addition to conserving energy resources and reducing energy costs for the consumer and the utility, the combined efforts through cooperation in conservation programs also have the potential to reduce program costs and increase participation in the programs (Dyballa and Connelly 1992).

One example of such cooperation is that of the Seattle Water Department and Seattle City Light, which blended their programs into a joint effort after independently planning to implement installation programs in apartment buildings (Dyballa and Connelly 1992). Puget Sound Power and Light, Metro (the regional sewer authority), and Washington Natural Gas collaborated with the Water Department and Seattle City Light to implement the joint utility program. The objective of the program was to deliver close to 800,000 residential retrofit kits that included showerheads, faucet aerators, toilet retrofit, and hot-water heater insulation wraps. Each utility was responsible for a designated element of the program and a certain share of the costs, but the Bonneville Power Administration provided an additional financial incentive by offering to pay for 75% of the retrofit program's costs (Dyballa and Connelly 1992).

Utility Information Programs and Campaigns

Utilities offer several different types of information programs. All of them have the objective of increasing energy efficiency through raising awareness about, and increasing knowledge of, energy efficiency potential, technology, and experience. Most of these types of programs target the entire population of customers and are only occasionally targeted to either the residential or commercial/industrial sectors. Informational programs instituted by utilities have included marketing campaigns, the establishment of information centers or services, the demonstration and evaluation of energy-efficient measures and technologies, the establishment of labeling and certification programs, telephone hot lines, and TV and radio spots. Few data exist supporting energy consumption savings associated with these programs.

Private-Sector Initiatives

The private sector has made important contributions to the development of energy efficiency programs and other initiatives that affect the multifamily housing sector. Rather than responding with stagnation to the restrictions placed on business opportunities by the energy crisis of 1973 and the ensuing focus on energy conservation and energy efficiency, the private sector has responded in many cases by interpreting the situation as an opportunity for innovation. Among the most important private-sector contributions are the expansion of the energy conservation services industry and the development of energy efficiency building standards.

Energy Service Companies

Energy service companies (ESCOs) have become significant contributors to the energy conservation movement by providing innovative energy conservation services to the government, utilities, and private sector. ESCOs have researched and developed new opportunities and initiatives for the conservation of energy and the implementation of energy efficiency programs and technologies. Many ESCOs use "energy performance contracting," a shared-savings arrangement between the owner and the ESCO in which the ESCO guarantees the owner a certain level of energy savings and shares a portion of the energy savings. (See Chapter 5 for a more detailed discussion of energy performance contracting.)

Traditionally, ESCOs have delivered services primarily to the commercial and industrial sectors. Early efforts to market energy conservation as a shared-savings investment for the single-family residential market failed as the up-front marketing and administrative costs outweighed the energy savings potential. But ESCO interest in the residential multifamily sector is on the increase, in part because of federal incentives recently made available to public housing authorities to enter into energy performance contracts.

One of the first ESCOs to begin serving the multifamily sector was Citizens Conservation Corporation (CCC), a Boston-based nonprofit organization formed in 1981 to augment the fuel assistance efforts of its parent company, Citizens Energy. CCC works solely in public and publicly assisted apartment housing, offering a comprehensive one-stop shopping approach that includes an audit, specification of energy-saving measures, financial arrangements, installation of measures, resident education and maintenance training, and monitoring of savings. CCC's goal has been to attract and invest capital to enhance both the energy efficiency and the long-term affordability of low- and moderate-income housing. In March 1995, many of CCC's building-based operations were acquired by Eastern Utilities Associates (EUA) Cogenex Corporation, forming a new, for-profit entity called EUA/Citizens Conservation Services, Inc., based in Lowell, Massachusetts. CCC remains as a nonprofit energy service organization focusing on research, consulting, and advocacy while continuing to do some building-based work.

In addition to its own direct energy service efforts, CCC has worked with DOE and HUD to help train public housing authorities regarding new HUD regulations that create an incentive for PHAs to invest in energy efficiency through performance contracts. To date, regional training sessions have been held in Boston, San Francisco, and Chicago. DOE provided funding for the initial sessions, and HUD has committed funds for future training sessions in other regions.

In Minnesota, the Northern States Power Company and the city of St. Paul established in 1981 a private, nonprofit ESCO called the Energy Resource Center (ERC) (Griffin et al. 1984). Now known as the Energy and Environment Resource Center (EERC), this organization develops and implements energy conservation projects for single-family and apartment residences and provides a full range of services in addition to program development, including both preliminary and detailed audits, design help for work plans, construction supervision, monitoring, and final inspection of the project.

Other ESCOs active in the multifamily sector include the Vermont Energy Investment Corporation; Wisconsin Energy Conservation Corporation; Conserve, Inc., in New York City; and the Syracuse Energy Services Company (SyrESCO) in Syracuse, New York.

Building Performance Standards.

The American Society of Heating, Refrigerating and Air-Conditioning Engineers (ASHRAE) is a private-sector organization that has made a significant impact on the development of energy efficiency initiatives across the United States. ASHRAE developed a voluntary energy standard for new buildings long before the federal government was able to develop mandatory standards. In 1974, when the then National Bureau of Standards (NBS) published the *Design and Evaluation Criteria for Energy Conservation in New Buildings*, ASHRAE used this document as the basis of its national voluntary consensus standard. The standard, ASHRAE Standard 90-75, Energy Conservation in New Building Design, was approved in 1975 and takes a performance-based approach to establishing reasonable energy efficiency requirements for new residential construction. Since that time, the ASHRAE standards have been revised several times to adopt them into enforceable code language and to update the technological criteria included in the standards. ASHRAE standards provide a basis for some of the federal energy efficiency standards described earlier.

Program Evaluation

Most energy efficiency program evaluations to date have been conducted for single-family residential or commercial/industrial programs, with relatively few evaluations of multifamily programs. This situation is due to the small number of programs that specifically target apartment buildings and to the lack of sufficient resources and lack of attention to program evaluation. Yet the need for evaluations of multifamily programs is great. Anecdotal evidence suggests that many building managers and landlords are hesitant to have their buildings retrofitted because they are skeptical about whether the savings can be realized. Evaluations can provide testimony that energy efficiency improvements are worthwhile in apartment buildings.

Isolating the savings generated by any energy efficiency program is more complicated than simply comparing utility bills from before and after the retrofits. As the understanding of energy use and savings has become more sophisticated, evaluators have developed methods to control for changes in weather and other factors that could alter energy consumption but that are not due to the energy efficiency program. The process of multifamily program evaluation is more complicated than single-family program evaluation because the number of factors affecting consumption levels is much greater in apartment buildings. For example, the space conditioning and domestic hot-water distribution systems are more complex, air movements are difficult to trace, and resident behavior is more varied. In addition, vacancy and occupancy changes in apartment buildings, which can significantly impact energy consumption patterns, are difficult to analyze. These factors also make it difficult to conduct pre- and post- comparisons with occupancy and behavior held constant. The split incentives among building managers, owners, and tenants further complicate the process. Additional roadblocks to multifamily program evaluation may include multiple systems in a single building, accessing multiple billing meters, and legal definitions regarding when landlords must supply heat.

The development of many new evaluation methods has been driven by utility DSM programs, which have had to pass rigorous regulatory screenings to prove their effectiveness. These regulatory concerns have also forced evaluators to adjust savings for free-riders (those who would have achieved savings without program incentives) and free-drivers (those who achieved savings as a result of a program, but who are not directly subsidized). Although the majority of the program evaluation work does not directly address multifamily programs, many of the various methods and techniques developed in the non-multifamily sphere are applicable to multifamily programs.

Evaluation Needs of Different Program Types

Because program evaluation is an emerging field, a universally accepted method for calculating the savings generated by an apartment building energy efficiency program has not yet been established. In any case, each evaluation should be tailored to the individual program's needs and resources, which do tend to be similar within types of programs.

Public Programs

Because public programs tend to use limited resources to partially retrofit as many buildings as possible, use of in-depth evaluations on a building-by-building basis to determine total program success is usually prohibitively expensive and time-consuming. However, evaluations remain crucial tools for determining the effectiveness of programs and how they could be improved, as well as for justifying continued support of effective programs.

The evaluation of a public program usually focuses on billing data from participating buildings and a control group that has been selected to match the retrofitted buildings. Additionally, survey information is commonly used to supplement utility readings. Although surveys do not directly measure energy savings, they are useful tools in evaluating public reaction to a program and in identifying possible contributions to savings or the lack thereof.

Utility Programs

The most extensive evaluations of apartment building retrofits have been conducted by utility companies. These evaluations have set the few guidelines that exist for highly accurate evaluation of apartment building energy efficiency improvements. Stringent regulatory requirements force utilities to conduct comprehensive evaluations, which calculate not only the energy savings generated but also the cost-effectiveness of those savings and associated peak-load reductions. Some public utility commissions require more extensive analysis, including evaluation of the program cost-effectiveness using different economic variables to model various perspectives. A utility's task of multifamily program evaluation is facilitated by its access to billing records and by its ability to apply to the multifamily sector its experience in evaluating other DSM programs.

Description and Analysis of Evaluation Methodologies

Just as program type varies greatly according to several factors, such as funding source, building type, technical opportunities, and local market acceptance, so, too, do evaluation methodologies. Accordingly, there is no single best evaluation methodology. Table 4-2 summarizes the characteristics of the evaluation methodologies described below. Table 4-3 defines some of the terminology commonly used in evaluation of a building's energy use.

Billing/Statistical Analysis

The most commonly used method for evaluating apartment building energy efficiency programs is billing and statistical analysis. Basically, this method compares pre- and post-retrofit energy consumption while controlling for variables that could affect consumption. This

Table 4-2

Key Characteristics of Methodologies for Determining Building Energy Use

Tool	Cost	Accuracy	Ease of Use/ Timeliness	Reliability/ Knowledge of Errors	Complexity	Ability to Control for Nonprogram Effects
Billing/ statistical analysis	Low-moderate	Moderate	High	High	High	High
Instrument-based	High	High	Moderate	Moderate	Moderate	Low
Surveys	Moderate	Moderate	Moderate	Low	Low	High
Site visits	Low	High	Moderate-low	High	Low	Low
Combination of methods	Moderate	High	Low	High	High	High

approach yields a change in consumption that can be directly attributed to the program measures, as opposed to changes resulting from nonprogrammatic factors (for example, arbitrary resident behavior or changes in weather or energy price).

A major advantage of billing and statistical analysis is its relative ease of application. Although some statistical experience is necessary to conduct the evaluation, most studies do not use statistical methods more advanced than ordinary least-squares regressions. The statistical result also reports the accuracy of the estimated savings and a range in which the true value for savings is most likely to fall. Billing analysis is simpler than many other techniques and, after the data have been gathered, is easy to complete in a relatively short period of time. By incorporating heating degree days into the analysis, an evaluator can control for changes in weather. Adding a control group of similar but nonparticipating buildings also enables the evaluator to control for general trends due to changes in societal energy consumption or changes in price (Kushler et al. 1992).

Billing and statistical analysis is not without its limitations, however. Data collection and removal of outliers is very time-consuming, often more so than the actual analysis of the data. Furthermore, the soundness of this method depends upon the accuracy and amount of data collected. Billing analysis for only a few units is often not a statistically reliable indicator of the effect of an entire program. Moreover, identifying appropriate control buildings and collecting the necessary

Table 4-3
Commonly Used Evaluation Terms

Baseline consumption The annual amount of energy that would have been consumed had there been no energy efficiency program. Gross and net savings are calculated from this projected consumption.

Free-driver A person who undertakes the steps encouraged by a program because of learning about them through the program, but who does not actually receive the incentives offered.

Free-rider A person who receives the incentives of a program, but who would have undertaken some or all of the retrofits without the program. Free-ridership is less of an issue in apartment building energy efficiency programs than in other programs because few energy efficiency improvements would be undertaken in apartment buildings without incentives.

Gross savings The total change in energy consumption that occurs after a program has been completed. The measurement includes decreases or increases in consumption that would have occurred had the retrofits not been installed. Gross savings are frequently adjusted to control for changes in weather.

Heating degree day A unit of measurement representing the effect of temperature on heating fuel consumption. The number of heating degree days during one calendar day equals a building's balance-point temperature minus the average outdoor temperature of that day.

Impact evaluation The measurement of the effect that a program has on the amount or cost of energy consumed within a target group. Impact evaluation examines energy savings and peak-load reduction.

Net savings The change in energy consumption that can be attributed directly to the program. It is most often derived by subtracting from gross savings the change in consumption measured in a control group.

Process evaluation The evaluation of the qualitative aspects of operating a program. Process evaluation examines program design, delivery and operation, marketing issues, customer satisfaction, and contractor-customer issues.

Verification The process of confirming that energy efficiency measures have been installed as planned and continue to be used by the residents of the building.

Self-selection bias The effect of individuals with certain traits choosing to enter or not to enter a program. As a result, program participants may share common traits that nonparticipants lack, making comparison of savings between participants and nonparticipants difficult.

data from them can be a difficult process. Changes in occupancy within a master-metered building can also confound results. Although this factor can be controlled for, estimates of occupancy rates and patterns are often imprecise.

A commonly used billing analysis tool is PRISM (the Princeton Scorekeeping Method), a computerized evaluation tool created at Princeton University (Fels 1986). PRISM evaluates savings by analyzing weather-adjusted pre- and post-retrofit consumption data. Advantages of PRISM are that it automatically estimates a variable degree-day base and provides standard error estimates for all parameters. These error estimates are valuable for indicating how well the billing data validate the presumed relationship between energy use and outdoor temperature and for judging the statistical significance of changes in building performance.

A number of limitations and potential pitfalls are pertinent to PRISM and other billing data–based evaluations of apartment buildings. Any type of billing analysis is beset by data quality problems that often occur with apartment buildings. Such problems are particularly acute for oil-heated properties, where billing may not be automated and analysts are often confronted with handwritten records irregularly recorded. Fels and Reynolds (1993) developed techniques for improving data quality and procedures for applying PRISM to reliably analyze consumption in oil-heated buildings, such as those common in New York City. Another drawback of PRISM is that it cannot account for changes in occupancy rates, which are often an important factor in apartment buildings (Mills et al. 1987). Also, PRISM analysis on unit-level data can introduce error due to the thermal effects of mass and heat loss between units in a building. Building-level analysis is more appropriate for the multifamily sector because unit-level data do not vary independently (Okumo Tachibana 1994). Originally, PRISM was not able to analyze savings in buildings that heat and cool with the same fuel, but an enhancement to the tool—the development of the Heating-and-Cooling (HC) model—has cleared this obstacle (Fels et al. 1994). Nevertheless, there is little analytic experience in addressing the combined complications of apartment buildings plus heating-and-cooling operation. Despite the limitations, PRISM is a valuable tool for evaluating retrofit installations, particularly because of its strength in accounting for seasonal, weather-dependent variations in energy use.

End-Use Metering

End-use metering accurately measures pre- and post-retrofit consumption by directly measuring the energy usage of individual devices. This technique can be used to measure gross savings, evaluate individual retrofits, supplement billing analysis, and estimate the interactive effects of a combination of retrofits. The strength of end-use metering is that it provides a record of energy usage broken down by specific use. Unlike billing data analysis, which only provides a picture of how an entire building or apartment uses energy, end-use measurements reveal very specifically the amount of energy consumed by different devices. In addition, by measuring usage directly, the evaluator avoids many of the data collection problems of billing analysis.

The disadvantages of end-use meters have traditionally been their high cost, large physical size, and unattractive appearance. However, the assumption that submetering is prohibitively costly and that it necessitates expensive, bulky hardware is perhaps becoming obsolete. Thanks to advances in large-scale integrated circuit technology, data loggers with multiple channels for recording binary and analog data are becoming remarkably inexpensive and elegant (Kinney 1994).

One problem that remains with end-use metering is that, in order to measure change in consumption, the meters must be installed and operating for a period of time measuring pre-retrofit consumption. Delaying the retrofits to obtain pre-retrofit meter data delays the savings, which compounds the project cost and logistical difficulties (Kushler et al. 1992). Further information on end-use metering techniques is given in our discussion of audit tools in Appendix C.

Engineering Estimation

Although engineering estimates of program savings are primarily used as an audit tool, this technique can be a valuable complement to measurement-based evaluation approaches. With engineering estimation, the steady-state performance of old equipment is measured (or estimated), and the savings are calculated to equal the difference in efficiency between the old and the new devices. Engineering estimation approaches can range from simple algorithms to complex computer models. Generally, complex real-time modeling yields more accurate results than simple steady-state heat-loss algorithms. The main shortcoming of engineering estimation is its inaccuracy. The source of the inaccuracy is that engineering estimations account very poorly for human behavioral factors and application difficulties. Sophisticated building simulation models can more accurately account for some of these factors, but their expense renders them beyond the scope of many programs (Kushler et al. 1992). Finally, although engineering estimation can complement other evaluation techniques, it is no substitute for empirical methods.

Surveys and Site Visits

Surveys and site visits provide a less quantitative and more qualitative tool for evaluating the impact of a program. Surveys can be conducted in person, by telephone, or by mail and are designed to collect information from participants, affected nonparticipants, program staff, and others involved in the program. Site visits are used to verify that energy conservation measures were installed as planned and are still being properly used, and to complement other evaluation methods by collecting data and making observations that can be accomplished only on site. The importance of these two methods is that they include end user information beyond meter readings. Surveys can be used not only to estimate energy consumption patterns but also to assess the residents' opinion of a program. Frequent surveys can improve evaluations and reveal which energy efficiency improvements are most likely to be favored by residents (Cason et al. 1991). Site visits can provide an important gauge of changes occurring in an apartment building that may be affecting energy consumption. For example, a site visit evaluation conducted by the Bonneville Power Administration for one of its DSM programs revealed measure malfunctions that would not have been discovered otherwise (Hickman and Steele 1991).

The biggest weakness of surveys is that participants tend to answer questions in the manner in which they think the survey writer wants the question to be answered. Consequently, the results of survey collections tend to be skewed toward overestimating compliance with the new measures and reduction in energy-wasting habits. Additionally, surveys and site visits are used effectively only as a complement to other methods, thereby adding an additional expense to the program evaluation (Kushler et al. 1992).

Combining Methodologies

The most effective evaluation approach for a specific program is a combination of evaluation techniques that is financially feasible for the individual program and that will fulfill the program's evaluation requirements. Combining methodologies is one of the most significant trends in the evaluation field (Kushler et al. 1992). For example, a number of recently completed multifamily program evaluations utilized a combined methodology, complementing analysis of energy bills with engineering estimates, site visits, or survey information.

These evaluations use a combination method called "leveraging." Leveraging involves taking the results of one type of analysis and applying those results in another method, in order to make the second method more accurate than it would have been otherwise. Another combination technique is "triangulation." Triangulation estimates savings by comparing the results from two or more evaluation methods and selecting the most likely middle ground. Frequently, triangulation is used to calculate free-rider ratios and other factors that are difficult to measure accurately (Kushler et al. 1992). Triangulation can also be used to combine engineering estimates and billing analysis.

Complex building analysis, which is commonly used to evaluate savings for a performance contract, focuses analysis more specifically on how the residents of individual buildings consume energy and how their consumption changes rather than just how the amount billed changes. The analysis combines billing data with specific information about the efficiencies and usages of energy-consuming devices in the building. This method requires site visits or occupant surveys, combined with engineering estimates to complement billing analysis. Building analysis is most useful when it is necessary to make extremely accurate measurements of an individual building's changes in consumption.

Multifamily Program Evaluations

Although there have been few evaluations of apartment building programs, we present as case studies the findings of four organizations that have had significant multifamily program evaluation experience in the past ten years: Lawrence Berkeley Laboratory, Seattle City Light, Wisconsin Energy Conservation Corporation, and Citizens Conservation Corporation.

CASE STUDY
Lawrence Berkeley Laboratory

A team at Lawrence Berkeley Laboratory has compiled and analyzed data from energy efficiency projects in apartment buildings across the country. The effort, part of the Building Energy Use Compilation and Analysis (BECA) project, has collected information on apartment building retrofit projects from many different organizations, including utilities, government agencies, housing authorities, research institutions and laboratories, and energy service companies. Summary results from this sectoral evaluation are listed in Table 4-4.

Table 4-4

Summary of Savings and Economic Indicators from BECA Database on Energy Conservation Efforts in Multifamily Housing

	All Buildings	Fuel Heat (Privately Owned)	Fuel Heat (Public Housing)	Electric Heat
Number of retrofit projects	191	111	38	42
Energy savings[a] (MBtu/unit/yr)	9 ± 1	15 ± 2	12 ± 4	5 ± 1
Energy savings (%)	15 ± 1	16 ± 2	13 ± 2	14 ± 2
Retrofit cost (1987$/unit)	600 ± 100	260 ± 80	580 ± 220	1,600 ± 240
Investment intensity (years)	1.0 ± 0.2	0.4 ± 0.1	1.7 ± 0.3	3.4 ± 0.4
Payback time (years)	7 ± 1	4 ± 1	10 ± 0.4	23 ± 7
CCE[a] (1987$/MBtu)	5 ± 1	3 ± 1	8 ± 2	11 ± 3

Source: Adapted from Goldman et al. 1988b, 24 (Table 7).
Note: Values are medians ± standard errors.
[a] Electricity savings are converted to site MBtu using 3,413 Btu = 1 kWh.

The data are primarily from buildings with five or more units, about a third of which were built before 1940, and 80% of which are low-rise structures (Goldman et al. 1988b). Most of the buildings (72%) have central heat and are master metered, and gas is the most common heating fuel (55%). Another 22% are heated by electricity (typically, baseboard resistance heating), almost all of which are located in the Pacific Northwest. Oil-heated buildings account for 19%. The sample includes large numbers of buildings located in Minneapolis/St. Paul, New York, New Jersey, Philadelphia, Chicago, San Francisco, and the Pacific Northwest. For the most part, these areas represent regions where utilities or other organizations have played a leading role in developing multifamily retrofit programs.

In the data sample, retrofit efforts have tended to focus on reducing space heat and domestic hot-water energy use. Public housing retrofits were more likely to include window and boiler replacements. In fuel-heated buildings, heating system measures were the most common strategy, with heating controls installed in more than half the buildings and heating system retrofits in about a third. Attic insulation was installed in only about 20% of the fuel-heated buildings. Electrically heated buildings received mostly envelope measures and measures to reduce hot-water usage, including low-flow showerheads. About half of the electrically heated buildings received attic and floor insulation. Median retrofit costs were substantially lower in fuel-heated buildings ($370/unit) than in electrically heated buildings ($1,600/unit) (Greely and Goldman 1989). The evaluators attributed these large differences in cost to variations in retrofit type (system or shell) and differences in how comprehensive the program designs were. Additionally, evaluators found that electrically heated buildings, which were smaller on average, achieved economies of scale to a lesser degree than their larger, fuel-heated counterparts.

In most cases, the group used PRISM to analyze whole-building energy consumption data before and after the retrofit. The evaluators then analyzed the cost-effectiveness and relative magnitude of the conservation retrofit investment and compared the cost of conservation investments to purchases of fuel or electricity. Multivariate regression analysis was used to determine which characteristics had the most impact on the level of energy savings in the sample.

The study showed median annual energy consumption decreased by 14 MBtu/unit (16%) in fuel-heated buildings and 1,450 kWh/unit (14%) in electrically heated buildings (Goldman et al. 1988b). The evaluators found that the economics of retrofitting central heating systems in fuel-heated buildings were very attractive, with an average payback of four years for privately owned buildings. Older, fuel-heated buildings retrofitted with both heating system and shell measures achieved savings of 26% with payback periods of less than six years. By contrast, payback periods for electrically heated building retrofits were typically twice as long, and the evaluators' results indicated that expenditures greater than $2,000/unit were not cost-effective. The evaluators determined that the cost-effectiveness of electrically heated building retrofits could be improved by focusing on highly inefficient buildings and by designing less expensive retrofit projects that target lighting and domestic hot-water systems.

CASE STUDY
Seattle City Light

Seattle City Light's Multifamily Conservation Program entered its pilot phase in 1986, providing financial and technical help to owners of apartment buildings with five or more units and electric space heat. The buildings had two or three stories and were typically 20 to 30 years old. Most had electric baseboard space heat and individual electric water heaters. The retrofit measures, detailed in Table 4-5, included double-glazed windows, attic or roof insulation, under-floor insulation, wall insulation, caulking, weatherstripping, efficient-flow showerheads, pipe and duct wraps, additional cavity venting, and common-area lighting modifications. A program cost breakdown according to type of measure installed is shown in Table 4-6.

When Seattle City Light (SCL) first began to evaluate its multifamily program in 1989, there were no multifamily evaluations upon which to base a research design, except a few studies from the Northeast. The methodologies from these studies were inappropriate for SCL's application because of differences in climate and because of the prevalence of central heating systems in the other climate zones. Okumo (1991) had to develop her own methodological approach to analyze the program. She extensively evaluated the 1986 and 1987 pilot phase of the program and used the results to make predictions about the impact of continuing the program.

The evaluation, which was conducted over several years, relied primarily on billing and statistical analysis, supported by data from on-site audits and inspections. Data collected during site visits were used as input for some of the regression analyses. In conducting the analyses, Okumo evaluated the savings by both common-area usage and individual-unit usage for low-income buildings and standard-income buildings. She concluded that energy savings were about 10% for low-income apartments, about 14% for standard-income apartments, 21% for common areas in low-income buildings, and 36% for common areas in standard-income buildings (Okumo Tachibana 1994).

Table 4-5

Energy Conservation Measures Used in the Seattle City Light Multifamily Program (Percentage of Building Receiving Each Measure)

	Standard-Income Buildings		Low-Income Buildings		
Measures Installed, by Program Year	Treatment Cohort			Treatment Cohort	
	I 1986	II 1987	Ia 1986	Ib 1986	II 1987
Common-area lighting	9%	71%	0%	0%	38%
Windows	96%	93%	100%	91%	100%
Insulation	52%	66%	92%	64%	75%
In dwelling units					
Venting	52%	54%	92%	73%	68%
Pipe/duct	26%	32%	83%	64%	62%
Strip/caulk	13%	0%	42%	55%	84%
Other	9%	0%	0%	36%	19%
Showerheads	26%	71%	100%	0%	24%

Source: Adapted from Okumo 1991, 23 (Table 3).

Table 4-6

Percentage of Total Cost Attributable to Each Measure Category in Seattle City Light Multifamily Program

Measures Installed, by Program Year	Standard-Income Buildings		Low-Income Public Housing Buildings	Low-Income Privately Owned Buildings	
	Treatment Cohort		Treatment Cohort	Treatment Cohort	
	I	II	I	I	II
	1986	1987	1986	1986	1987
Common-area lighting	2%	8%	0%	0%	3%
Windows	87%	80%	75%	81%	78%
Insulation	6%	12%	16%	13%	12%
Measures in dwelling units	4%	1%	8%	6%	8%

Source: Adapted from Okumo 1991, 69 (Table 24).

Okumo extended the analysis, using linear regression incorporating analysis of covariance, to calculate the savings attributed to individual apartment retrofits across program years and regardless of tenant income level. (Individual apartment retrofits consisted of window, envelope, and hot-water measures.) Okumo used information from 1,449 retrofitted units in 111 buildings and a comparison group of 1,365 units in 95 buildings. She found that net savings in 1988, within a 95% confidence interval, were 1,050 kWh (±5%) per residential unit. By building type, the separate results were 1,082 kWh (±8%) for low-income units and 1,078 kWh (±4%) for standard-income units (Okumo 1991).

Okumo chose analysis of covariance among second-year participants as the best estimator of common-area savings. (Common-area savings were generated by energy-efficient lighting improvements.) Using information from buildings that have metered common areas, Okumo found that net savings in 1988 were 520 (±14%) kWh per residential unit (Okumo 1991). By building type, the separate results were 469 (±8%) kWh for low-income buildings and 595 (±9%) kWh for standard-income buildings (ibid.). Analysis of covariance was chosen as the most reliable estimator for common-area savings because it adjusted for different consumption in the pre-period, that is, for an imperfectly matched control group (Okumo Tachibana 1994).

These methods of analysis depend upon comparison with a control group of similar buildings. Okumo chose "pre-participants" from the program's waiting list to comprise the control group. This group most closely matched early participants in the program, and their change in consumption reflects the change in consumption that would have occurred in the retrofitted buildings. Selecting buildings from the waiting list as controls represents one alternative when evaluating an immature program for which short-term estimations of program savings are desired. At a later stage of program maturity, selection of a nonparticipant control group becomes more imperative (Okumo Tachibana 1994).

Lighting retrofits to common areas accounted for about 74% of electricity savings measured by building meters, with the other 26% of savings occurring in buildings that did not receive lighting retrofits (Okumo 1991). Replacement windows provided approximately 70% of the total tenant-area energy savings, the remainder being supplied by insulation and other envelope measures, along with efficient-flow showerheads. Replacement windows made up 80% of the total measure cost (ibid.).

Given the low cost of electricity in the Northwest (about $0.03/ kWh), SCL found that the savings generated in 1988 by standard-income participants in the pilot phase of the program were marginally cost-effective for the Bonneville Power Administration region, but not for the SCL service area. The savings from low-income participants were not cost-effective from either perspective. From a regional perspective, the loan program for standard-income participants was calculated to have yielded a positive net present value of $202,000, but the grant program for low-income participants had a negative net present value of $614,000 (SCL 1991a). The evaluation concluded that the difference was due primarily to the higher costs of the low-income portion, which required certification of tenant income, some repair costs, and non-energy-saving measures, but offered less flexibility in choosing retrofit measures and contractors (ibid.).

Although the SCL evaluation experience offers valuable lessons in an area in which information is sparse, the extent to which the results of the SCL program evaluation can be used as a model for other multifamily program evaluations is somewhat limited by the fact that the moderate marine climate of the Seattle area is unrepresentative of the rest of the country (Okumo Tachibana 1994).

CASE STUDY
The Peoples Gas Light and Coke Company

In 1991, the Wisconsin Energy Conservation Corporation (WECC) completed an evaluation of the Peoples Gas Light and Coke Company's Energy Savers Fund program (Graham et al. 1991). The program, which was carried out by the Center for Neighborhood Technology and other community-based organizations in Chicago, provided low-interest loans to owners of apartment buildings between 1984 and 1989. The program was discontinued in 1989, in part because it was found to be not cost-effective under the criteria specified for such utility programs in Illinois.

The loans issued under the Energy Savers Fund program could be used to pay for approved energy conservation measures (ECMs) and some rehabilitation and repair expenses. Through the program, owners of 316 apartment buildings (comprising 7,500 units) received $9 million worth of financing. The apartment buildings were typically brick walkup, 80 to 100 years old, with 10 to 20 units each. Most buildings had gas-fueled single-pipe steam heating systems. The measures installed in the 65 evaluated buildings were storm and replacement windows, indoor thermostats, heating system repairs and boiler replacements, radiator work, and ceiling insulation (Table 4-7).

For their evaluation, WECC relied primarily on billing data analysis and evaluated the program using several criteria: total energy savings generated by the program, the cost-effectiveness of the program, the frequency and costs of individual retrofits, and the impact and cost-effectiveness of each ECM. Using an analysis that considered a control group of similar buildings, the study concluded that the buildings participating in the program reduced energy consumption by 238–511 therms, representing a 20–37% drop in pre-retrofit consumption. Without comparison to the control group, the estimated savings were

Table 4-7

Energy Savings from Apartment Building Retrofits Performed by the Peoples Energy Savers Fund in Chicago

Energy Conservation Measures	Number of Installations	Share of Program Cost[a]	Average Lifetime (Years)	Estimated Energy Savings (Therms/Unit)[c]	Estimated Benefit/Cost Ratio[c]
1. Heating system replacement	29	23%	19.5	373 ±169	1.58 ± 0.50
2. Heating system retrofits	45	7%	7.6	81 ±106	1.30 ± 1.44
3. DHW system retrofits	28	4%	7.1	64 ±111	1.25 ± 1.83
4. Insulation	33	7%	22.5	207 ±150	4.43 ± 2.45
5. Storm windows	40		10.0	81 ± 85	0.37 ± 0.28
6. Replacement windows	15	47%[b]	23.8	174 ±171	0.72 ± 0.43
Measures 1–4 combined	18	—	21.0	458 ±134	1.55 ± 0.61

Source: Based on Graham et al. 1991, 412 (Table 1) and Pigg 1994.

[a] Other measures not separately evaluated, including lighting retrofits and infiltration reduction, accounted for the remaining 12% of program costs.

[b] Cost breakdown not available for storm versus replacement windows, so a cost share is given for both of these measures combined.

[c] Estimates developed from regression analysis, with 95% confidence limits showns as ± values.

188–420 therms per unit, or 17–39% of pre-retrofit usage (Graham et al. 1991).

To evaluate cost-effectiveness, WECC calculated benefit/cost ratios (BCRs) from three perspectives: total resource cost (by using a test approved by the Illinois Commerce Commission), building-specific, and participant-specific. The total resource cost test yielded a BCR of 0.77, indicating that the program was not cost-effective. From other perspectives, however, the program was cost-effective, with a building-specific BCR of 1.77 and a participant BCR of 1.80 (Graham et al. 1991). Costs were analyzed by measure category, with windows and heating system replacements constituting 70% of all measure expenditures (Table 4-7).

Finally, Graham et al. (1991) analyzed the energy savings that could be attributed to each type of retrofit. The effect of some measures on gas consumption was found to be not statistically significant, as indicated by the large uncertainty intervals shown in Table 4-7. The study also did a cross-sectional analysis of gas savings by building type and found that the "very large buildings showed significantly lower per-unit energy use, ECM expenditures, and energy savings...." (ibid., 414). Additionally, the inclusion of large buildings in the total resource cost test skewed the results. When applied only to the five very large buildings sampled, the total resource cost BCR was 0.41. Removing these buildings from the sample of 65 yielded a total resource cost BCR of 0.90 for the remaining 60 buildings. (A program breaks even when its BCR equals 1.0. A BCR of less than 1.0 indicates that costs were greater than benefits.)

CASE STUDY
Citizens Conservation Corporation

Citizens Conservation Corporation (CCC), the Boston-based non-profit energy service company discussed in Chapter 3, has used energy performance contracts to complete retrofits of 30 public and publicly assisted multifamily developments over the past decade. The buildings CCC has treated have typically been large, master-metered buildings with very high energy use. Most have been low-rise buildings built in the 1940s and 1950s, with central oil- or gas-fueled steam heating, although CCC has also treated electrically heated high-rise buildings.

Many of the buildings contained inefficient, oversized boilers with no controls, and almost all buildings suffered from excessive air leakage, stemming from low-quality, or lack of, insulation in attics and walls, as well as poorly sealed windows, doors, and vents. Most of the boilers were in the second half of their expected lifespans. CCC's approach to retrofitting such apartment buildings has emphasized a "building-as-a-whole" approach, and in most cases, equipment measures such as boiler replacement and controls have formed a key component of this integrated energy efficiency strategy. Temperature-limiting thermostats or zone valves, envelope improvements, and water-saving measures have also been typical components to CCC's retrofit packages.

CCC monitors the savings at each project using a proprietary computerized billing analysis program that adjusts for weather effects. The billing analysis is supported by site visits. The monitored savings provide a valuable record of what works and what does not work in this type of housing stock and contributes to the ongoing discussion of the persistence of savings in low-income multifamily housing. In a study funded under the DOE-HUD Initiative, CCC recently evaluated the persistence of savings in the gas-heated buildings it has treated for which at least three years' worth of monitored savings records were

Figure 4-2

Persistence of Energy Savings for CCC Apartment Building Retrofits in New England

Plotted by number of years of post-retrofit data

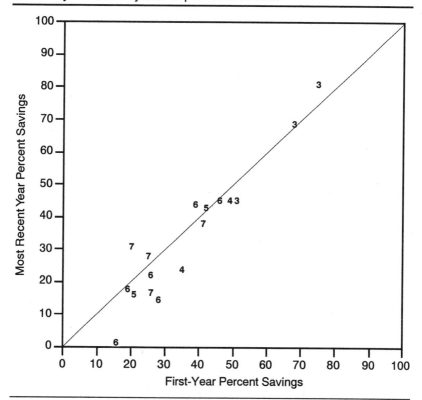

Source: Based on Nolden and Snell 1995, 6 (Table 2).

available (Nolden and Snell 1995). The sample is comprised of 17 gas-heated public and publicly assisted housing developments in New England, which were retrofitted in 1985–1990. At 15 of the developments, the retrofit included replacement of the boilers with highly efficient units; the other 2 retrofits included replacement of the domestic hot-water heaters. The monitored savings show that savings can, and often do, persist in projects involving equipment measures.

Figure 4-2 plots the most recent year measured energy savings against Year-1 savings for the 17 projects evaluated by Nolden and Snell (1995). Monitored energy savings ranged from 5% to 80%, averaging 35% and matching the 35% average projected savings, which ranged from 14% to 60%. With regard to savings persistence, the study found that on average, the savings did persist beyond the first year. An average of five years of post-retrofit data was available for the sample. For the first criterion of savings persistence—the average Post-Year-1 savings compared with Year-1 savings—the mean for the 17 developments was 91%. For the second criterion—the most recent year's savings as a percentage of Year-1 savings—the mean was 89%.

Several mechanical, structural, and behavioral factors affect the persistence of savings. In particular, the study found that savings persistence depends on the delicate balance of three key factors: appropriate measure selection, proper maintenance of equipment, and resident behavior. Careful attention to each of these issues is crucial in ensuring savings persistence. The mechanism of performance contracting provides an ESCO with a major incentive to make sure that these factors are properly addressed. In CCC's experience, performance contracting has fostered a long-term relationship between CCC and the building owner or public housing authority, whereby CCC can provide ongoing technical support, monitoring of savings, and resident energy education.

Conclusion

Despite the usefulness and value of the above evaluations and similar studies, the relative lack of information about the effectiveness of multifamily programs remains a salient issue. Additionally, evaluations from one geographic region are often of little use in other geographic regions because of differences in climate, age and type of building stock, fuel usage, heating system type, and utility DSM incentives. Clearly what we need are more evaluations that focus on specific geographic areas, heating fuel and system type, building size, building construction material, and ownership patterns.

Chapter Five

Financing Energy Efficiency Improvements in Apartment Buildings

The ability of apartment building owners to obtain financing for energy efficiency improvements depends on a variety of factors, many of which have little to do with the building's actual need for improvements or existing level of energy efficiency. The creditworthiness of the owner; type of public assistance, if any, available to the property; location of the building; current debt-to-value ratios and debt ceiling limitations; tenant income levels, rent receipts, cash flow, and vacancy rates; tax issues—all of these contribute to both the willingness and the ability of the owner of an apartment building to undertake energy efficiency improvements. Moreover, the problem of split incentives between owners and tenants, discussed in previous chapters, serves to inhibit energy conservation investments in apartment buildings.

Chapter 2 documented the reality of income distribution among residents of multifamily housing: only 8% of households living in apartment buildings have incomes greater than $50,000. Not only are the apartment dwellings inhabited by higher proportions of low-income households, but also the overall number of available apartment units continues to decline. The *State of the Nation's Housing* report notes that housing inventory removals from 1984 to 1989 included almost 200,000 rental units per year (JCHS 1993). By roughly 2005, on the order of a half-million apartment dwelling units may be lost as a result of expiring federal mortgage subsidies and accompanying tenant income eligibility restrictions, and the continuing disinvestment or substantially lowered investment by public and private investors in multifamily building

175

stock (National Low Income Housing Preservation Commission 1988). Investments and incentives to invest in maintaining, rehabilitating, and improving the energy efficiency of this stock have not kept pace sufficiently to prevent a continuing decline in multifamily housing conditions. Moreover, cash flow problems are common in the multifamily sector. About 25% of properties with five or more units experience negative cash flows, although the incidence drops to 11% for larger properties of 200 or more units (Fronczeck and Savage 1995).

In this chapter, we look at the barriers associated with financing energy improvements in multifamily housing, review types of public and private investment mechanisms (including grant and loan funds), and identify some successful strategies that have been tested by various entities over the last ten years. Our discussion is organized under three categories: (1) public housing; (2) publicly assisted, privately owned housing; and (3) nonassisted privately owned housing. Within these categories, it is also important to note the distinction between master-metered and individually metered dwellings, as the financial incentives for the owner and the tenant must be differently structured to attract investment.

Types of Financing for Multifamily Energy Conservation

The types of financing available for apartment building energy efficiency investment vary largely depending on the characterization of the property. For public housing, funds throughout the 1980s were primarily provided by the U.S. Department of Housing and Urban Development (HUD) through modernization and other capital improvement grants. Only in the early 1990s were regulatory incentives created allowing public housing authorities to use debt financing for these improvements.

In publicly assisted, privately owned housing, development tax incentives and energy tax credits were available during the early 1980s, in addition to second mortgages, refinancing, and other traditional bank financing mechanisms. Accessing these limited funds or credits, and convincing owners to apply capital solely for energy improvements, is not a simple undertaking.

Finally, in purely privately owned apartment dwellings, owners may be reluctant to undertake additional debt for energy conservation improvements when faced with other capital improvement needs, even when the efficiency improvements can decrease overall operating expenses and save the owner money over the long term. Owners may be at the limit of their debt capacity or may simply perceive the borrowing process as an additional complication.

In order to discuss the specific financing approaches for energy efficiency improvements available to the different types of multifamily housing, commonly used financing terms, such as *grants, loans, bonds, tax credits, leases, lease-purchases*, and *energy performance contracting*, need specific definition.

Grants

Grant programs make money available for improvements or equipment purchases without an accompanying expectation of repayment. For instance, in public housing, modernization grants from HUD are made available to housing authorities for making structural repairs or capital improvements or installing energy efficiency improvements. Grants are usually the preferred method of financing energy assistance for low-income households, whether in the form of cash assistance for utility costs (as described under the LIHEAP program in Chapter 4) or through payment for labor and materials to install energy efficiency improvements (as in the U.S. Department of Energy Weatherization Assistance Program [WAP]). Because low-income households often carry a higher burden of energy costs in relation to income, as described in Chapter 2, and are also not in a financial position to repay a loan, grants are often provided by both government and utility programs to achieve energy efficiency for these households. In a successful program, providers of grant monies also benefit through reduced arrearages, reduced need for fuel assistance subsidies, and reduced peak consumption.

Loans

Any time funds are borrowed, whether from a bank or other source, the money is considered loaned. This type of financing mechanism is also termed *debt financing*. There are many different types of debt financing mechanisms, but generally all require some form of security or collateral from the borrower. Factors affecting the ability of a borrower to finance energy improvements through loans include creditworthiness of the borrower; amount of equity or security available from cash, property, or land; debt coverage ability (that is, how much debt is allowed by the financing entity in relation to property cash flow); and loan-to-value or debt/equity ratio (the ratio of the debt to the property value) (Weedall et al. 1986). Loans are sensitive to term, interest, and transaction costs.

Weedall et al. (1986) describe two types of debt financing. The traditional approach is recourse financing, in which the borrower assumes all of the risk of the debt but also receives all of the benefits. In the second approach, project financing, a lender evaluates the potential cash flow of a project for the anticipated repayment stream. Several investors may contribute to a project financing, but usually a higher interest rate is required to hedge the risk associated with this "off-balance-sheet" financing approach (ibid.). The return in off-balance-sheet financing is also usually spread among several investors, not just the borrower.

Bonds

Another form of debt financing is the issuance of bonds, or pledges to repay numerous investors a determined interest rate over a fixed period of time. Many public entities, such as municipalities, states, and authorities, as well as private entities such as utilities and corporations, can issue bonds. Those issued by public entities are usually tax-exempt and therefore provide a lower interest rate. Utilities and corporations may issue bonds for a particular and costly (usually more than $1 million) purpose, such as new utility generation capacity.

Weedall et al. (1986) describe the two types of bonds—general-purpose and revenue—as being similar to traditional and project financing. General-purpose bonds are backed by the credit standing of the entity issuing the bonds (usually rated by a credit rating agency); revenue bonds are backed by the anticipated project-specific revenue and usually are more risky and more costly than general-purpose bonds. Weedall et al. note that energy efficiency improvements have rarely been financed through bonds, primarily because of the relatively small size of the investments and the difficulty in identifying the risk of achieving energy savings as predicted.

Tax Credits

The 1978 Energy Tax Act, passed as part of the National Energy Conservation Policy Act (NECPA), provided three major tax incentives to stimulate residential and business energy efficiency. These incentives included a maximum residential energy tax credit of $300 for installation of certain qualified energy improvements (identified as 15% of the first $2,000 spent on energy efficiency); a credit of up to $2,200 for the installation of qualified solar and wind energy property; and a business energy tax credit of 10% of the cost of qualified improvements, with no dollar limitation (Keppler 1978). In the early 1980s, apartment owners were able to take advantage of the business energy tax credit until it was phased out in 1986. Courts became active in citing developers for use of these tax credits as "abusive tax shelters" (Garland 1986) by overinflating the cost and depreciation of energy and alternative energy source tax credits.

Holly Courts, a public housing complex in San Francisco, was the site of third-party-financed energy performance contract for solar domestic water heaters.

In 1990 Congress passed the Low Income Housing Preservation and Resident Homeownership Act (LIHPRHA). Coupled with tax credits for development of low-income properties, LIHPRHA offers potential for rehabilitation, including energy conservation, of multi-family properties in the publicly assisted, privately owned sector.

The LIHPRHA legislation was intended to help sustain the continued affordability of properties originally developed under various HUD incentive programs during the 1960s and 1970s. At that time, in return for the incentives, developers agreed to certain mortgage restrictions, including reservation of units for low-income households. These buildings are also called *expiring use restriction properties* because the original 30-year mortgage lock-in on the income restrictions is expected to expire in the 1990s. Because these properties were considered to be at risk of sale to private-market owners (and subsequently, to private-market rents), the LIHPRHA legislation was passed in an effort to provide certain rehabilitation and preservation incentives to facilitate nonprofit or tenant ownership and prevent resale to the open

179

market. Administered by HUD, the LIHPRHA program offers federal subsidies for nonprofit ownership so that little cash is required for acquiring the property. The combination of tax credits for low-income housing development and the LIHPRHA legislation may contribute to renewed interest in reducing overall capital expenditures on these properties through energy conservation improvements (see discussion of expiring use restriction properties under Publicly Assisted, Privately Owned Housing, below).

Leases and Lease-Purchases

Weedall et al. (1986) note that the Internal Revenue Service (IRS) rules regarding leases and lease-purchase agreements are constantly changing. There are, however, certain tax advantages to true leases, and certainly risk reduction advantages to the owner of an apartment building. Also, in a true-lease situation, risk of equipment failure or poor performance is transferred entirely to the supplier. Therefore, leasing equipment for energy efficiency, particularly in the context of an energy performance contract, shared or guaranteed savings plan, or other structure, is attractive to some building owners. Additionally, if financial constraints are present, leasing equipment provides an attractive alternative to a loan as there is little up-front capital required on the part of the owner.

Installment contracts or lease-purchasing agreements, on the other hand, eventually transfer ownership of the property to the lessee. Depending on the ownership of the equipment and the provisions of the lease, certain tax incentives may be available to either the lessor or lessee. Under the IRS code, only removable equipment that is not a structural part of the building may be considered for a true lease. Although it is not within the scope of this book to identify all of the tax issues associated with acquisition and depreciation of energy-related equipment, suffice it to say that these issues can be complicated and are frequently changing according to tax law and court decisions. Accelerated depreciation, interest deduction, investment tax credits, and other issues affect the equipment owner's tax situation and do serve to make energy equipment ownership more attractive.

Energy Performance Contracting

Energy performance contracting refers to an arrangement between an owner and an energy service company (ESCO), in which the ESCO conducts an energy analysis of the property, projects the savings to be achieved as a result of certain energy efficiency improvements, may arrange the financing for the improvements, and, depending on the how the project is structured, may guarantee that the savings will be achieved or may take its fee from those savings. The financial technique associated with energy performance contracting is that the ESCO uses the energy savings to help amortize the cost of the improvements, or the equipment lease, or the installment/service contract. ESCOs use a variety of financing techniques to provide that the performance of the energy measures will contribute enough capital over time to retire the debt, or at a minimum to ensure that the owner will pay no more for the improvements than would have been paid in the absence of the measures. ESCOs use a number of different risk reduction mechanisms to ensure that the energy savings will accrue as predicted (see Performance Contracting Approaches, below).

Barriers to Energy Efficiency Investment in Apartment Buildings

Financing energy efficiency improvements in apartment buildings is affected by a large number of variables, some of which were mentioned earlier in this chapter and are related to the general barriers to energy efficiency described in Chapter 1. A key factor affecting an owner's willingness or ability to invest in energy efficiency is the availability of capital to make the improvements. Many times, the capital constraints of an owner drive the decision-making process as to which measures will be installed. Sometimes the lack of available capital leads an owner to invest in short-term, quick-payback items while sacrificing the potential to undertake longer-term improvements that would deliver much greater savings over time.

Competing capital needs also affect an owner's ability to undertake energy efficiency improvements. When many improvements need to be made on a building and capital is limited, energy efficiency improvements may not take the highest priority, even though the energy savings could help pay for the measures over time. In public housing, where capital improvement needs far outweigh available grant funding for modernization, there is a constant struggle to prioritize capital versus energy needs. In publicly assisted, privately owned housing, the need to provide capital for lead paint abatement in older properties often supersedes the installment of energy efficiency measures. In private housing, the perceived "hassle factor" of obtaining additional debt contributes to an owner's unwillingness to install additional energy efficiency improvements (Colton et al. 1994).

Debt ceiling limitations, loan-to-value ratios, and creditwor-thiness also limit an owner's ability to invest in energy efficiency improvements. When an owner is cash strapped, or the rental income from a property is limited, additional debt cannot reasonably be repaid from the owner's operating capital; therefore, lenders are reluctant to provide additional financing. An owner's own credit rating with the bank, and potentially a property's location in a low-income area, may result in denial of traditional bank financing for a project. Finally, the amount of total debt compared with the value of the property, expected to fall below 75–80% of the property value in traditional lending, may prohibit further investment. A lender's appraisal of the property, particularly if energy investments are undertaken, may result in an assigned value that is higher than the value attributed to the neighborhood, resulting in an "overimproved" property—and denial of a subsequent request for financing. This barrier has been noted in numerous affordable housing conferences. Later in this chapter we discuss methods of mitigating these barriers.

Security and collateral requirements of lenders may also present a barrier to investing in energy efficiency improvements. This issue becomes apparent in many energy performance contracts, in which the energy service company anticipates using a portion of the savings achieved to help amortize the debt. Because a lender cannot quantify to a degree of certainty that the savings will be sufficient to retire the debt, as well as provide positive cash flow while the improvements are undertaken, a loan provision is frequently denied. **Underwriting procedures** of typical lenders, while taking into account the cash flow of the property, do not typically account for the reduced utility costs accompanying energy conservation activity and therefore are not adjusted to take into account anticipated cash-flow benefits from such improvements.

Lack of appropriate information regarding energy efficiency is another barrier to apartment building energy efficiency financing. Many owners recognize that outdated heating systems may need repair but are unsure of how to proceed. As a result of utility demand-side management (DSM) efforts, owners may be aware that an energy audit is offered, but, as noted in Chapters 3 and 4, the targeting and participation rates of apartment dwellings in these efforts have been limited. Therefore, the amount of technical information available to owners and particularly to lenders is limited. Payback criteria and methodologies must be carefully explained to both owners and lenders. Utility financial incentives available to the owner are usually based on specific avoided-cost criteria; what the owner views as desirable may conflict with what the utility can offer as an incentive. For example, an owner may be interested in making capital improvements, such as window replacement, that add aesthetic or even real value to the property but which are difficult to justify on the basis of energy savings. As noted by Stephen Strahs, "Unless the *current* cash flow situation is an acceptable basis upon which to make the loan, the bank must be persuaded that implementation of the improvements will result in an operating margin improved substantially by the resulting energy cost savings" (Strahs 1981). Traditional lenders simply are unprepared to deal with the technical complexities of financing large-scale energy conservation and are thus apt to assign a higher risk value to the financing package.

Finally, the **willingness of the owner** to undertake the perceived burden of additional financing, coupled with a lack of knowledge about the long-term advantages of investing in energy efficiency improvements, must be taken into account when financing energy conservation in apartment buildings. Later in this chapter, as we examine some of the successful financing strategies available to this market, we will see that all such strategies entail a considerable **education activity** required to simultaneously move the owner and the lender to invest in conservation measures.

Financing Mechanisms for Energy Improvements in Apartments

Overcoming the barriers to financing efficiency improvements in the multifamily sector requires creativity and determination. No one approach is appropriate for all situations. For public housing, tools such as the updated Performance Funding System are now in place that hold promise for expanding the rate of energy efficiency investments in this important housing category. Opportunities to better address the publicly assisted, privately owned sector exist through both recent programs established to preserve affordable housing as well as innovative use of older programs. Leadership by state and local organizations, such as those in Vermont and the Minnesota Twin Cities, have demonstrated effective methods for reaching private apartment building owners. The reach of all of these efforts is enhanced when traditional leaders realize the value of energy efficiency investments in helping to secure better cash flow for any property, as witnessed by the use of energy-efficient mortgage programs. Finally, performance contracting can enlist entrepreneurship in helping to provide both energy efficiency investments and follow-through with efficient operations for all categories of apartment housing.

Public Housing

HUD spends roughly $3 billion annually on energy bills for publicly assisted households, including direct reimbursements for public housing authority (PHA) utility expenses as well as utility allowances for residents in public and assisted housing.* Since the 1970s, energy efficiency in public housing has been primarily funded through two HUD mechanisms—operating funds and modernization/capital improvement funds. As described in Chapter 4, several HUD modernization grant programs provide funds that can be applied to energy efficiency improvements. Spending on energy efficiency upgrades was relatively high in the early 1980s, when many PHAs undertook energy conservation measures with their modernization grant funds. Compared with the 1977 level, HUD's overall modernization budget declined 86% by 1988, and the number of new or rehabilitated units declined 91% (Green and Grande 1992). Spending on energy efficiency dropped disproportionately and now represents a smaller share of the shrunken supply of modernization funds.

Due to decreased funding, many housing authorities are forced to use their limited funds to support maintenance and repair in an effort to prevent an overall decline of their properties. Energy efficiency improvements have not been a priority for most PHAs. Unfortunately, much of the nation's central-city public housing stock, as a result of this decreased funding, has fallen into the "distressed" category. Consequently, it has been proposed that PHAs be authorized to use their comprehensive grant funds for unit replacement, not simply rehabilitation, which would be a major public housing policy shift.

The Performance Funding System

In 1975, HUD created the Performance Funding System (PFS), a mechanism by which HUD provides operating capital to PHAs. Under PFS, PHAs were allowed to retain 50% of any savings attributable to energy efficiency measures, with the remaining 50% going back to HUD. The amount of savings is based on the use of a three-year rolling base of consumption. As consumption levels increase or decrease during this period, HUD adjusts its payment to the PHA, allowing the PHA to retain as an incentive 50% of any energy savings. However, as was noted by many at the time, this structure failed to

---

* Data specifically documenting federal expenditures on energy bills for public and publicly assisted housing are not readily available; we derived a $3.1 billion estimate from RECS (1990b, Table 18), which is roughly consistent with a $3.5 billion estimate provided by Groberg (1995).

provide PHAs a sufficient incentive to reduce energy consumption, for two major reasons. First, HUD basically guaranteed that the operating costs for utilities (including sewer and water) of a PHA would be funded annually as part of the PHA's operating budget, and to the extent a PHA was able to save energy, its overall operating budget was, in effect, reduced under this policy (Ferrey 1986). Second, PHAs that wished to make major energy efficiency improvements were unable to access enough capital to do so (Mills et al. 1986).

Changes in the Performance Funding System enacted through the Housing and Community Development Act of 1987 and implemented in a final rule in 1991 created financial mechanisms to enable PHAs to access private capital to finance energy efficiency improvements (ORNL 1992). PHAs were then able to access debt financing from banks, utilities, and energy service companies under performance-based agreements for a maximum contract period of 12 years. Under the former 50% rule, the PHA was only able to recoup 1.5 times the value of 1 year of savings over a 4-year period. Under the 1991 regulations, a PHA may retain 100% of the savings for the duration of the performance contract, so long as at least 50% of those savings are used to retire the debt (Manheimer 1992). About a dozen public housing authorities, including such major authorities as Chicago, have engaged energy service companies, but it is too early to have statistical results regarding the energy savings attributable to these projects.

The 1991 rule authorizing debt financing through energy performance contracts, and the authorization for PHAs to leverage their modernization funds for debt financing, will increase the amount of private as well as public capital used in public housing to finance energy and other improvements. Several types of debt financing are now available to PHAs for energy efficiency improvements, including loans, tax-exempt bond issuances, leases, and lease-purchase arrangements—any one of which may be partially or wholly repaid with the energy savings achieved after installation of the improvements.

One of the case studies discussed below identifies how this new regulation can be used to increase private capital investment in energy efficiency improvements in public housing, as well as to blend energy efficiency improvements with overall capital needs to provide a comprehensive capital improvement project that is more attractive to the housing authority.

A complicating factor is the utility allowance system, which provides an allowance for energy costs to public housing residents who pay for their own utilities. This factor affects publicly assisted, privately owned housing as well. Utility allowances are structured so that a rent adjustment is made for tenants who, by statute, pay no more than 30% of their income for rent but who are burdened by the need to pay for some or all utility expenses. The allowances provide a credit to the tenant for the projected utility cost, which is supposed to be determined annually by the PHA or owner. This credit is then applied directly to the tenant in the form of a deduction in the percentage of income applied to the rent payment. The difference between the amount the tenant pays and the amount of the actual rent is then provided directly by HUD to the PHA or property owner (Lubke 1994). This arrangement costs the federal government more than $1 billion annually and offers little cost control incentive, either to tenants, whose rent payments are being reduced, or to landlords, who are receiving the difference directly from HUD. The utility allowance system is one of the most difficult situations in public and assisted housing in which to determine adequate financial incentives to foster energy conservation improvements.

Publicly Assisted, Privately Owned Housing

Much of the nation's apartment housing falls into the category of publicly assisted, privately owned housing. Such units encompass properties for which the owners received certain public subsidies, usually in the form of mortgage insurance or interest write-downs from HUD, during the 1960s and 1970s. Many of these properties were built in central cities of the Northeast, Midwest, and West Coast. Included in the publicly assisted category are properties served by HUD's Section 236, Section 221(d)(3), and Section 8 new-construction and rehabilitation programs. Originally designed to increase the supply of affordable rental housing for low-income households, the HUD programs often provided financial incentives in return for developers and owners guaranteeing that a certain number of units are occupied by low-income households. A recent article published by the National Assisted Housing Managers Association enumerates the assisted housing programs particularly well (Lines and O'Brien 1994).

According to statistics summarized by Lines and O'Brien (1994), there are 6,037 older (early 1960s) publicly assisted properties housing 674,227 families; additionally, there are 4,154 newer (1970s) assisted properties housing 361,882 households. Of the older dwellings, 59% are considered to be in the "distressed" category, based on a variety of factors, including cash flow, vacancy rates, need for capital improvements, and age of dwelling. Of the newer properties, 13% are considered distressed. Lines and O'Brien (ibid.) estimated that an unfunded needs backlog of $955 million exists, representing about $4,000 per unit in distressed properties. When combined with the ten-year decline in HUD's budget for assistance through modernization and capital improvement funds, the lack of available capital to assist these properties is clearly a major problem.

Many of these HUD-assisted or HUD-insured properties fall into the category of expiring use restriction properties. As noted earlier in the chapter, these properties are now reaching the point at which the owners' original mortgage restrictions have expired, pushing the properties toward possible resale at market prices and the loss accompanying affordability provisions. The 1990 LIHPRHA legislation allows for certain tenant protections and encourages purchase by nonprofit or tenant-controlled entities as a means of protecting the affordable nature of these expiring use restriction properties.

In publicly assisted, privately owned properties, tenant rents and the HUD subsidies represent the two main revenue sources for operating costs and capital improvements. In at least two states, Massachusetts and New York, substantial state subsidies also support privately owned apartment properties. HUD requires that all of its assisted properties maintain reserve accounts for funding repairs and replacements. However, only 45% of properties have reserves sufficient to cover backlogged needs (Lines and O'Brien 1994).

Private owners of assisted housing have been able to tap into a variety of sources over the past ten years for energy conservation improvements because of their wider ability to access debt financing. However, owners may still be inhibited by limitations in available programs, lender criteria, and information. The bulk of assistance to this category of apartment housing has come from a combination of grant and loan programs provided through HUD, including the Community Development Block Grants (CDBG), Urban Development Action Grants (UDAG), and Flexible Subsidy Programs. In terms of motivating energy conservation improvements, a number of HUD programs specify energy efficiency standards for some rehabilitation work; however, use of these guidelines has not been tracked or enforced, and in particular, no energy efficiency standards are attached to work done using CDBG funds. Some utility assistance has been available, but apartment housing participation in and incentives provided by utility DSM have been limited. During the 1980s, some states set up innovative apartment building conservation financing through the use of oil overcharge funds.

The ability of states and cities to use these rehabilitation and development monies to provide innovative financing techniques as part of program design resulted in significant apartment building conservation activity in some areas during the early 1980s. The most successful programs provided grant/loan blends or interest write-downs, coupled with active intervention in the debt process. For instance, under the DOE pilot program begun in 1986 for conservation of apartment buildings, the state of Washington used its $100,000 portion to match owner contributions for conservation, thereby increasing the leveraging power of weatherization grant funds (Callaway and Lee 1988). The HUD Solar and Energy Conservation Bank allowed for such a grant/loan leverage capacity; however, despite an initial authorization for $3 billion over a four-year period, actual appropriated funds totaled only $65.9 million between 1981 and 1986, and the program expired in 1987. Of this amount, about $5.8 million was spent on about 5,300 conservation projects in the multifamily sector (HUD 1986).

In 1984, the city of Chicago, collaborating with neighborhood organizations, including the Community Investment Corporation and the Center for Neighborhood Technology, established the Chicago Energy Savers Fund through a combination of HUD CDBG ($5 million) and utility-provided funds ($10 million from Peoples Gas Light and Coke, financed through a surcharge on customers' utility bills). This fund was established with several goals in mind, including job creation, energy conservation, capacity building of local community groups, and preservation of affordable housing. The project was designed to stabilize housing, using a relatively low per-unit cost of $1,500/unit. A major feature of this project addressed the underwriting problems discussed above and included projected energy savings (30% maximum) in determining debt coverage requirements (Freedberg and Schumm 1986). Through the program, owners of 316 apartment buildings (comprising 7,500 units) received $9 million worth of low-interest financing. Energy savings from these projects averaged 22% (Graham et al. 1991). As described in Chapter 4, the program was discontinued in 1989 because it was deemed not cost-effective under the state-mandated total resource cost test.

The city of Portland, Oregon, also undertook an innovative approach to the use of HUD CDBG and UDAG funds for energy conservation. In the mid-1980s, the city allocated $3 million in UDAG monies to leverage $12.5 million in private capital (Hemphill 1981). Both by providing capital and by changing local CDBG guidelines to mandate the provision of certain energy efficiency measures as a precursor to rehabilitation, the city was able to include energy efficiency investments in its overall housing redevelopment strategy.

Many cities across the country undertook innovative grant/loan packages through CDBG and UDAG programs throughout the early 1980s. However, with the general decline in HUD's budget (see under

Public Housing, above), CDBG and UDAG funds also declined. UDAG monies were phased out entirely in 1993.

In addition to the uses of CDBG and UDAG monies to leverage private capital for conservation, many states undertook innovative projects through the use of oil overcharge funds and unique blends of weatherization grant funds. Two states in particular, Massachusetts and New York, developed innovative financing for apartment building conservation using oil overcharge dollars in combination with other grant programs, such as WAP or LIHEAP. These funds were primarily targeted to the publicly assisted, privately owned apartment housing sector, although in some instances both states selectively used WAP grant funds to leverage private or other public investment in publicly subsidized housing (both states are unique in that they operate both federally and state subsidized public housing units).

In Massachusetts, between 1984 and 1990, $10 million in oil overcharge funds was used to target expiring use restriction properties with conservation dollars. This initiative, the Expiring Use Restriction Weatherization Assistance Program (EURWAP), assisted over 2,000 of the 5,000 units in this category in the state, using a combination of grants and loans.*

Frequently, the developments were also being brought under new ownership by nonprofit or tenant groups, as was the case with a 500-unit property located in South Boston; a 300-unit property located in Somerville, Massachusetts; and a 200-unit property located in Lowell, Massachusetts. In these instances, the EURWAP grants and loans were often used in combination with tax credits as part of a development "work-out" package in which the investors sought to rehabilitate properties in conjunction with new ownership. Through the unique combination of grants, low-interest loans, and tax credits, at least 800 units of affordable housing were prevented from being lost to market-rate rents. The energy conservation funds were used to make additional capital improvements, including heating system work and replacement windows and doors. Through the accompanying reduction in operating costs, estimated to range between 15% and 25% or higher, the properties were able to maintain healthy cash flows to support debt coverage by the new owners. The program also served as a model for development of a similar program for expiring use restriction properties in North Carolina.

---

* This program was designed by the Massachusetts Executive Office of Communities and Development by one of the authors (DeBarros); hence, most of the information in this section derives from direct experience.

In New York, weatherization funds and oil overcharge funds supported the development of a nonprofit agency whose sole purpose is to obtain private financing for conservation in low-income apartment properties. The agency, Conserve, Inc., draws on a combination of public and private resources and serves as a financial packager for apartment properties housing low-income eligible households. A particular function of the agency is to help arrange owner matches to WAP funds and to assist in obtaining loans for owners who may otherwise be considered marginal for debt financing by a traditional lender. This interface with lending institutions relies on an analysis of the building's pre- and post-conservation cash flows to convince lenders to provide capital or to increase the amount of debt coverage authorized. A further description of Conserve and some of the conservation financing issues it tackles is provided in the case study at the end of this chapter.

The role of Housing Finance Agencies (HFAs) across the nation in providing private capital for rehabilitation and energy efficiency is critical (Sachs and Hunt 1992). HFAs are the mortgage holders for a major portion of publicly assisted, privately owned apartment dwellings in the country. Because of their ability to raise capital through the issuance of both tax-exempt and taxable bonds, HFAs can offer a number of financing techniques to this market segment. Allan Hunt, of the Vermont Housing Finance Agency, notes that by providing more flexible lending criteria and often serving as the lenders of last resort, HFAs are in a unique position to require that certain energy efficiency standards be met when development, refinancing, or rehabilitation occurs, as well as to provide capital for that purpose (Hunt 1993). Vermont, Massachusetts, New York, Rhode Island, Connecticut, Michigan, and Minnesota are several states having HFAs that have developed energy efficiency packages along with housing development or rehabilitation efforts. By allowing for flexibility in determining debt coverage amounts, loan-to-value ratios, and underwriting criteria, HFAs can increase the amount of private capital available for conservation and package financing for owners of apartment properties who might not otherwise be approved by traditional lenders for debt financing. Many HFAs also allocate development tax credits and are in a position to require that certain energy efficiency requirements be met as a condition to receiving tax credit allocations. In Massachusetts and Vermont, HFAs have successfully used energy performance contracting as a financing technique. These projects are discussed in the case studies at the end of this chapter.

Four additional financial strategies may be used in the future to support energy conservation in public and assisted housing: the Low

Income Housing Preservation and Resident Homeownership Act of 1990 (LIHPRHA), the Comprehensive Housing Affordability Strategy of 1990 (CHAS), the Low Income Housing Tax Credit, and the Community Reinvestment Act (CRA) of 1977.

As mentioned earlier, prior to 1990 there were few federal protections available to assisted housing developments in the expiring use restriction category. With the passage of LIHPRHA in 1990, certain tenant protections and financial incentives were provided to enable the transfer of ownership of these properties to nonprofit agencies and to the tenants themselves. By encouraging the inclusion of energy conservation measures in the overall financing package, as demonstrated in Massachusetts and North Carolina, future operating costs for utilities can be lowered substantially so that tenant or nonprofit ownership becomes a feasible long-term strategy.

The 1990 Cranston-Gonzalez Housing Act also set up a new state and local funding strategy entitled the Comprehensive Housing Affordability Strategy (CHAS). For states and local entities to receive HUD dollars for housing initiatives, a five-year affordable housing strategy must be filed (Haber 1994). States and local entities are required to hold public hearings on the CHAS, and energy conservation can be built into the program design and financing techniques planned for these strategies. The CHAS must also detail how other housing funds, including funds from the Farmers Home Administration and the Housing Finance Agencies will be utilized and coordinated.

The Low Income Housing Tax Credit, which continues to be reauthorized at the federal level, allows for certain development tax incentives to be available for new construction and substantial rehabilitation of low-income housing. It allows for funds to be provided annually by a syndicate of investors "to support the development and cover the operating debt service" for a specific time period (Haber 1994). States are required to hold public hearings on their plans for the allocation and use of tax credits. This process could also allow energy conservation standards to be included within the application materials and plan for awarding tax credits to developers.

The 1977 Community Reinvestment Act (CRA) was passed as part of banking legislation to stimulate lending and financial servicing in low-income communities. New regulations governing these requirements are being promulgated, and many housing development agencies and advocates are actively seeking stringent standards for investment in affordable housing and economic development in these neighborhoods. Conserve, Inc. (see case study, below) has promoted the use of energy performance financing as part of the overall development strategy for reinvesting in low-income housing in central cities, including using traditional bank financing under CRA as the primary financing for such projects (Woolams 1994).

Many states have now created Affordable Housing Trust Funds to assist in the financing of low- and moderate-income housing. Haber notes that, in the last ten years, 25 housing trust funds have been created at the state level, 9 at the county level, and 32 at the municipal level. These trust funds, as is the case with the Vermont Housing Trust Fund, can often be accessed for affordable housing with fewer credit and debt restrictions than may be applicable with traditional lenders. The Vermont Energy Investment Corporation and the Vermont Housing Finance Agency have worked closely to ensure that energy conservation measures are included within the program design and financing provisions of the housing trust fund in that state.

Private Housing

Government regulations do not exist to require or motivate energy efficiency in purely privately owned apartment buildings. On the other hand, depending on the perceived "bankability" of the owner, traditional debt financing, energy performance contracting, and some utility incentives are easier to access for privately owned apartment buildings.

The energy-efficient mortgage has been used with some success in Vermont at the time of resale or refinancing of a property. Whereas the national model for the energy-efficient mortgages through Fannie Mae and Freddie Mac* has been primarily targeted to single-family residences and has met with limited success, in Vermont the energy-efficient mortgage combines two elements—a uniform home energy rating system (HERS) and a mortgage process (Sachs and Hunt 1992). The Vermont Energy Investment Corporation notes that, by combining these elements, the energy-efficient mortgage allows for a greater debt-to-income ratio than in traditional lending and also provides a benefit similar to a decrease of 1.5% in the interest rate (ibid.). Since 1988, this program has resulted in 600 energy-efficient mortgages and includes over 15 lenders who represent 80% of the mortgage lending in the state.

Energy performance contracting has been used successfully by energy service companies (ESCOs) in providing leased equipment or shared savings for small commercial and residential apartment buildings. Many ESCOs that work primarily in the commercial sector are able to access utility incentives for lighting, motors, and other electrical

---

* Fannie Mae was once known as the Federal National Mortgage Association, but the official name is now simply "Fannie Mae." Freddie Mac is a common name for the Federal Home Loan Mortgage Corporation.

efficiency measures for small commercial-rate apartment buildings, focusing primarily on the common areas for lighting improvements.

With the recent attention on retail wheeling and the concept of electricity as a commodity, James Newcomb, president of E Source, writes in a May 1994 article that a new class of "super ESCOs" may well emerge that offers mass customization of service packages, including innovative pricing, power quality management, billing, and financial risk management (Newcomb 1994). These ESCOs, Newcomb notes, will have the capacity to leverage not only technical resources but also financial resources across large market segments.

Another model for financing energy efficiency improvement in privately owned multifamily housing is the extensive apartment building retrofit program carried out in Minnesota by the Center for Energy and Environment (CEE, formerly the Minneapolis Energy Office—see case study in Chapter 4). CEE has provided "one-stop shopping" for audits, retrofit planning, and financing low-interest loans. It acts as a loan agent to arrange financing for both energy efficiency retrofits and general rehabilitation of rental properties (CEE 1995). A Residential Loan Fund, derived from oil overcharge revenues and sponsored by the Minnesota Department of Public Service, provides short-term loans (limited to $7,000 in 1995) at below-market rates specifically for energy-related improvements. A more comprehensive Rental Rehab Loan program sponsored by the Minnesota Housing Finance Agency provides longer-term loans at low interest rates (up to $40,000 per structure or $8,000 per apartment in 1995). CEE assists in audit and retrofit planning and develops financing contracts with building owners. CEE's technical staff can ensure sound energy-saving projects, and owners can obtain financing for a set of retrofits tailored to a building and flexibly drawn from a set of efficiency improvements that are known to be workable and cost-effective. In some cases, low-cost retrofits are identified that building owners or property managers pay for themselves, without financing. CEE can also work out packages that blend energy retrofits with other sources, such as general rehabilitation financing. In this way, there are opportunities to "slip" the energy work, which may hold low owner interest per se, into an attractive package that meets a building owner's need for more comprehensive property improvements.

Performance Contracting Approaches

Energy performance contracting can be an appropriate way to manage financing for all categories—public, assisted, and private—of multifamily housing. Nevertheless, performance contracting carries several financial risks: (1) that the energy savings will not accrue as predicted and, if relied upon to help amortize the debt for installation, will result in default or delayed payment; (2) that the equipment installed will have operational or maintenance problems, resulting in higher expenses that detract from the projected savings; and (3) that the owner or ESCO will not remain solvent. All of these risks result in uncertainty about the ESCO's ability to make debt payments or achieve enough savings over time to reduce operating costs and provide positive cash flow to the building. These same risks occur regardless of the type of apartment dwelling affected by the performance contract (that is, public, publicly assisted, or private).

The risks of energy performance contracting can be reduced in a variety of ways. A performance guarantee can be structured out of pooled owner contributions, utility funds, or other public or private funds to reduce the risk of nonpayment or nonaccrual of savings. Energy savings insurance can be purchased that, although expensive, can greatly mitigate the risk that the predicted energy savings fail to materialize. Assignment language and contractual protections can hedge against the risk of ESCO insolvency or price fluctuations. Behavior and maintenance issues can be addressed through the provision of a long-term energy service contract covering maintenance staff training and resident education. All of these techniques can be used to make an energy performance contracting package more attractive to a lender.

In some cases, energy performance contracts can be administered as part of utility DSM programs. For example, SyrESCO, a nonprofit ESCO based in Syracuse, New York, has successfully operated a utility performance contracting program for low- and moderate-income individually metered apartment dwellings. Through a contract with Niagara Mohawk Power Corporation, SyrESCO is paid by the utility if measured savings reach a certain percentage of projected savings. For the contract amount, SyrESCO provides the technical assessment, installation of measures, monitoring of performance, and resident education required to achieve the savings.

Fannie Mae announced in mid-1993 a new utility energy efficiency financing program called REEP—the Residential Energy Efficiency Program. This program is beginning its first phase through an apartment building program sponsored by Pacific Gas and Electric Company (PG&E) as part of a pilot DSM bidding program. Fannie Mae will finance up to $3.5 million of energy efficiency investments in apartment buildings (public and private) through Citizens Conservation Corporation (CCC) acting as PG&E's subcontractor. The future demand-side revenue stream, which PG&E will pay to CCC for the projected 4,500 MWh of annual energy savings plus 1.5 MW of peak load reduction, will serve as security for the loan. Fannie Mae anticipates that this financing technique will increase the level of utility-sponsored energy conservation financing for apartment and other residential housing while improving overall housing affordability and preventing cross-subsidization in utility programs. Fannie Mae, PG&E, and the state of California have teamed up to offer zero-interest financing for measures to save electricity, gas, or water in apartment buildings. Initially, a $500,000 pilot project will blend PG&E and LIHEAP funds to hire ESCOs to provide audit, retrofit planning, vendor contracting, and installation supervision services to apartment buildings serving low-income households. The PG&E and LIHEAP funds will be used to write down the interest rate on retrofit financing, and Fannie Mae will purchase the loan from PG&E after completion of the project. Fannie Mae is also working with an energy financing company to introduce a loan tailored for energy efficiency improvements in HUD-assisted buildings; this concept involves finding a way to streamline HUD approval procedures to facilitate performance contracting and other debt financing instruments that are now inhibited by requirements that owners of HUD-assisted buildings obtain HUD approval to enter into debt arrangements.

Successful Energy Efficiency Financing Strategies

Of particular note for their innovative financing techniques are three nonprofit agencies that have targeted apartment building energy efficiency for at least ten years: the Vermont Energy Investment Corporation, Citizens Conservation Corporation in Boston, and Conserve, Inc., in New York City. Each of these agencies uses a slightly different financing technique, or combination of techniques, to stimulate private investment in energy efficiency improvements in all types of apartment housing.

CASE STUDY

Vermont Energy Investment Corporation

The Burlington-based Vermont Energy Investment Corporation (VEIC) was formed in 1986 through efforts of the Vermont Housing Finance Agency (VHFA) and others to fill a need for certain energy services statewide. At the time, the VHFA recognized the importance of controlling energy costs in its properties and the link between energy costs and over-all housing affordability. The VHFA took action to require that all apartment housing financed through the HFA meet energy efficiency standards and that underwriting procedures of the agency in financing both single-family and apartment properties should take into account energy costs in assigning a percentage of income to mortgage support. Effectively, this latter decision shifted the traditional underwriting criteria so that families could afford a higher debt. Recognizing that much of the state's apartment housing, particularly publicly assisted, privately owned housing, was electrically heated, the VHFA launched several joint initiatives in innovative financing with the Vermont Energy Investment Corporation.*

One of the VHFA's first actions was to identify the highest energy use properties and to provide financing for energy improvements by requiring that owners establish a Project Cost Escrow account, averaging $1,200 per unit, from which repairs and rehabilitation would be made. This account was formed at the initial time of the mortgage approval by applying a percentage of the total project cost. Decisions regarding the use of these monies had to be made jointly by the VHFA and the owner. Funds from these accounts were made available for more expensive efficiency measures, such as heating system conversions, along with reduced interest expenses resulting from refinancing properties from higher, early-1980s interest rates to lower rates. VHFA also offered to increase the owner's equity position in the project as a result of making the improvements.

VEIC provides the energy analysis expertise, an estimate of energy savings to be achieved, and, in some cases, a shared savings approach for apartment properties. Fuel switching of the electrically heated properties is a focus for the shared savings arrangements. In two apartment developments, the investment levels were approximately $80,000 to $90,000 for the fuel switch, with a simple payback of six years for one of the projects. VEIC and the VHFA also worked jointly to finance energy improvements for an expiring use restriction property, in which the energy savings assisted the tenant purchase of the property. For public housing developments, VEIC has successfully used VHFA financing to provide shared savings arrangements under the pre-1991 HUD regulations through the use of HUD waivers to allow for the preimprovement energy costs to be carried through the term of the loan (ten years). In one public housing authority project in Burlington, a cogeneration system was installed in a 160-unit high-rise development at a cost of $110,000, with a projected average annual savings of $25,000.

The unique partnership between VEIC and the VHFA allows for several innovations that are not traditional to lending institutions: (1) the VHFA can rely on the technical expertise of VEIC to identify the energy costs and potential savings, a relationship that allows a level of

---

* Information for this case study was provided through conversations with Beth Sachs and Allan Hunt as well as from their article (Sachs and Hunt 1992).

confidence on the part of the VHFA to offer more flexible credit policies and debt coverage allowances; (2) as mortgage holder to the bulk of apartment properties in the state, the VHFA can exercise its authority to require that energy efficiency standards be met or maintained at several crucial entry points to the mortgage process—sale, resale, or refinancing; and (3) the VHFA can provide nontraditional financing to such entities as public housing authorities and rely on VEIC's energy savings projections to provide the security for the projects.

CASE STUDY
Citizens Conservation Corporation

The programmatic approach and energy savings results of Boston-based Citizens Conservation Corporation (CCC) were introduced earlier in Chapters 3 and 4. Since 1981, CCC has provided over $12 million in energy financing through performance contracting in public and publicly assisted housing. In early 1995, the bulk of CCC's performance contracts were transferred to a newly formed for-profit ESCO, Eastern Utilities Associates/Citizens Conservation Services (EUA/CCS). CCC, which retains its nonprofit status, continues to serve the portfolio of the Massachusetts Housing Finance Agency as well as to provide research, fee-based consulting, and foundation-funded advocacy.

CCC's pioneering work in shared savings contracts was with public housing authorities in the towns of Brockton and Lawrence, Massachusetts. These pilot efforts were conducted prior to the 1991 HUD regulations and provided much of the basis for HUD's updated approach to energy performance contracting in public housing, particularly regarding HUD's approval to freeze the rolling base period under which energy costs are determined for purposes of energy performance contracting for a maximum period of 12 years. Three pilot projects in 1985 and 1987 undertaken by CCC with HUD waiver approval allowed for energy-based financing to occur in which the energy savings stream repaid much of the installation debt. In the Lawrence PHA, now approaching its tenth year of energy savings, almost $1 million in energy savings has been achieved over the life of a 12-year loan.

In addition to public housing work, CCC has also undertaken energy performance contracts in publicly assisted, privately owned housing units. Much of this work has been conducted through the Massachusetts Housing Finance Agency (MHFA), which has underwritten the interest rate for ten-year energy loans on MHFA properties. These loans are an extension of the first mortgage on the property and do not trigger a rent increase; rather, they are repaid from the energy and water savings, coupled with any additional grants, utility DSM, or other financing sources, including owner cost-shares. Most recently, CCC has participated as MHFA's contractor in the Massachusetts Expiring

Use Restriction Weatherization Assistance Program (EURWAP) discussed earlier.

CCC has tapped a variety of different financial sources to structure its energy performance contracts. In some instances, such as the Bay State Gas Program, a utility has provided an additional financial incentive for certain measures, such as an interest write-down or a principal guarantee. In public housing, CCC has been able to leverage both public and private dollars to support the energy performance contracts, using the public dollars for both grants and interest write-downs in order to blend and therefore reduce the overall interest rate for the project. Oil overcharge dollars have been tapped to provide low-interest capital and grant funds as well. By combining available utility, public, and private sources of capital to bring to a project, CCC has usually been able to secure below-market interest rates for its borrowers. Besides creatively blending various financial resources, CCC has worked to provide comprehensive and building-specific packages that allow both energy efficiency enhancements and needed capital improvements.

As oil overcharge funds dry up, federal resources decrease, and utility funds become less of a certainty, private capital will play an increasingly important role in financing energy performance contracts. In 1995, Chicago-based ChiCorp Financial Services, Inc., financed a $2.1 million performance contract for energy efficiency improvements in public housing managed by the Fall River Housing Authority in Massachusetts. This project, which was initiated by CCC but transferred to EUA/CCS in 1995, marks the first time that private capital has been used to fully fund an energy performance contract in public housing.

CASE STUDY
Conserve

As introduced earlier, New York City–based Conserve, Inc., is a nonprofit agency that was founded in 1984 through the support of the New York Department of State Weatherization Assistance Program and its subgrantees.* The organization's main activity is helping owners of apartment properties eligible for DOE weatherization assistance to access private capital for additional work, thereby providing a cost-share to the DOE WAP work. Conserve also assists local weatherization subgrantees in attracting owner interest and additional investment. In most cases, the properties served are privately owned and receive no public assistance.

As part of Conserve's approach, an owner receives an economic,

---

* Much of the information regarding Conserve is from conversations with the agency's executive director, Jack Woolams, as well as from marketing and other informational materials provided by Conserve.

energy investment, and loan feasibility analysis. The company examines the operating cost savings expected to accrue from energy work and packages this analysis in such a way as to attract a lender to a building that might not have received financing prior to the analysis. Conserve attempts to project cash flow sufficient for loan repayment and averages about a 30% internal rate of return on its projects. Using the WAP grants as leveraging power, Conserve packages smaller-scale ($10,000–$50,000) energy loans than traditional rehabilitation financing. Conserve's energy loans average $2,500 per apartment and, with cofunding from the WAP program, approach $4,000 per apartment. In 1992 and 1993, Conserve packaged loans for 134 apartment buildings, working with 22 weatherization agencies within the New York City area. These smaller loan sizes, and the accompanying cash-flow projections based on reduced energy operating expenses, lead a lender to provide the capital for the project, as Conserve walks both the owner and the lender through the financial analysis. Conserve gives special attention to marginal loans—those loans that can arguably be shown to be affordable given the reduced energy operating costs projected for the building.

In the last year, Conserve has piloted the concept of performance financing—an approach that differs from energy performance contracting in that there is no savings guarantee or shared-savings plan. Instead, Conserve develops a pre- and post-retrofit performance profile for a building, analyzing current and projected cash flows and operating costs, to convince a lender that the project is "bankable." Thus, in pursuing performance financing, the goal is to demonstrate that the operating cost savings from an energy project are greater than the loan payments, so that energy financing improves rather than burdens a building's cash flow (Woolams 1994).

Conclusion

Access to capital is a critical factor influencing apartment building owners' energy decisions. Already burdened by debt and, at least in the case of public and publicly assisted housing, limited to the amount of rent that is available for energy improvements, apartment building owners are reluctant to take on additional debt or may be deemed credit risks by traditional lenders. Add to these limitations competing capital needs, and the owner is even more reluctant to take out additional debt for energy improvements.

Successful energy efficiency financing strategies for apartment buildings share several components. First, many strategies combine public and private dollars to lower both the financial risk and the amount of owner capital required. Second, principal guarantees provided by utility or other public funds help hedge the risk associated with using energy savings to retire the debt for the improvements. Third, flexible underwriting procedures and credit policy approaches, backed by technical expertise, can release private capital to this market sector. Fourth, utility programs that specifically target incentives to apartment buildings and allow for owner cost-shares are likely to increase their apartment building participation rate. Finally, energy performance contracting and other shared-savings arrangements provide attractive packages in which risk sharing and blended capital sources can be linked to a project.

Note how all three case studies presented here involve adjustments to traditional lending policies. Conventional loan underwriting procedures do not include a review of energy costs and potential for conservation, both financially and technically. In the Vermont case, for example, the Vermont Housing Finance Agency adjusted its underwriting procedures to allow both for greater debt coverage ratio as well as for greater owner equity as a result of energy efficiency improvements. These policies allow the lender (VHFA) to provide more dollars to a project than would be provided under conventional financing. This process differs from traditional lending in two key ways: first, the technical capacity developed through VEIC provides the lender with documentation for energy costs and savings; second, VHFA can alter its policies to allow for a higher debt coverage or loan-to-value ratio. Most traditional lenders, working in a tighter regulatory environment, do not yet have this capacity or flexibility.

Through the Community Reinvestment Act and the recent energy efficiency investment activity of Fannie Mae, some of these credit policy, debt coverage, and loan-to-value adjustments may be achieved. Additionally, through the proactive setting of energy standards in the tax credit, LIHPRHA, and CHAS state plans or workout guidelines, states and local entities can actively require that development and rehabilitation initiatives include energy efficiency improvements. Finally, through the creative use of several financing sources and techniques, public-private partnerships can be created that foster energy efficiency improvements and provide the financing capacity to leverage private investment in apartment buildings.

Chapter Six

Recommendations

A number of suggestions for policies, programs, and research have been noted in previous chapters. In this final chapter, we synthesize these recommendations and present additional suggestions for actions that can be taken to stimulate, help finance, and successfully accomplish energy efficiency improvements in apartment buildings. Not all our recommendations have direct links to what was reported in the earlier chapters. Rather, we have drawn on our own experience and broader review work as well as the wisdom of others in the field to provide comprehensive recommendations for advancing multifamily building energy efficiency.

Considering the diversity of the multifamily sector in terms of physical building type, fuel use, geography, type of ownership, and the variety of institutions involved, a set of recommendations could be organized in a number of ways. We present recommendations by major player—federal agencies, state and local agencies, and utilities—followed by a separate section addressing research needs. Our closing thoughts point to the ultimate need for energy efficiency performance codes and standards to ensure that cost-effective investments in energy efficiency eventually reach the entire multifamily sector.

Making the investments needed to upgrade energy efficiency in apartment buildings requires substantial resources. Estimates of the sector's conservation potential reviewed in Chapter 2 suggest achievable site energy savings averaging 22% by pursuing appropriately intensive retrofit packages. However, Chapters 3 and 4 showed how much higher savings have been achieved when targeting buildings with the greatest need, such as those having much worse than average efficiency. The estimated total investment level to obtain substantial sectorwide savings is roughly $40 billion (1995$, updated from estimates of Goldman et al. 1988b). Thus, an investment of $4 billion per year would be needed to improve the multifamily sector over a 10-year time frame, corresponding to an average of $2,700 per dwelling unit. Clearly, government and utility programs cannot provide such resources, so these limited sources must be used to leverage large private investments. Chapter 5 highlighted some of the creative approaches that have succeeded in financing apartment building retrofit projects, demonstrating leveraging ratios of 20:1 or more.

Unfortunately, multifamily residential units remain underrepresented in government-funded and utility-sponsored conservation programs. In terms of resources, federal programs are still disproportionately focused on single-family and owner-occupied housing. State and local conservation programs have not met their potential for targeting the multifamily sector. Utility-based multifamily programs remain few in number and limited in scope. A greater commitment of resources to multifamily conservation is needed on all fronts. A number of good models now exist for programs that cost-effectively conserve energy in this sector.

Whether run by government agencies, utilities, or nonprofits, multifamily conservation programs should include the key elements summarized in Table 6-1. These are provision of objective and trustworthy information; targeted program marketing; financial incentives; technical assistance in retrofit planning (including oversight of bidding and contractor work) using the "building-as-a-system" approach; thorough energy and cost analysis; education of tenants, building staff, and management; and monitoring, follow-up, and evaluation. Targeting is important because of the diversity of the multifamily stock: different approaches are needed for publicly owned, publicly assisted, and private buildings; for large and small buildings; by fuel type; and according to the financial resources available. Knowing the right audience to reach is also important: sometimes it will be residents, but often it will be building owners and property managers, or even local contractors—the latter can be an important conduit for marketing energy efficiency programs to the privately owned apartment sector. Within a given market, it is valuable for a program to set up a single point of contact for coordinating the various services, providing "one-stop shopping" for building owners. Finally, performance standards, codes, and other regulatory tools are an important complement that should be considered by governments (local, state, and federal) and as a condition for

---

Table 6-1

Elements of Success for Achieving Energy Efficiency Improvements in Apartment Buildings

---

- Provision of information about opportunities

- Targeted program marketing

- Financial incentives

- Technical assistance, including contractor training and oversight

- "Building-as-a-system" approach to retrofit planning and implementation

- Thorough cost/benefit and cash-flow analyses

- Education of residents, management, and staff

- Performance monitoring, retrofit follow-up, and program evaluation

Various actors may emphasize different elements. But everyone should be aware that all pieces must come together to achieve widespread and reliable energy savings in the multifamily sector. Program initiators (federal, state, local governments and utilities) should backstop implementation efforts using regulatory tools or contingencies, such as codes and standards for building, equipment, and retrofit performance.

---

public and utility financing of retrofits. Such a comprehensive approach is needed to ensure significant and persistent energy savings.

Federal Programs and Policy

Federal programs addressing energy use in apartment buildings, at both the Department of Energy (DOE) and the Department of Housing and Urban Development (HUD), are seriously underfunded and inadequate to ensure that economic investments are made in efficiency improvement. A stronger federal role is needed, per our recommendations in Table 6-2. The barriers confronting energy efficiency in this sector explain the failure of market forces to beneficially impact multifamily housing in most of the country. These barriers (see Table 1-1) also imply that substantial and creative government involvement is needed to catalyze cost-effective investments. To date, DOE's programs—oriented to product R&D, standards setting, weatherization grants, and information provision—have not had the impact in the multifamily sector that similar activities have achieved in the single-family sector. Reasons for the relative lack of success include low

---

Table 6-2

**Recommendations for Federal Policies to Advance
Energy Efficiency in Apartment Buildings**

- Increase funding for programs targeting apartment buildings, dedicating at least $20 million per year (DOE plus HUD).

- Use federal grants to leverage other funding to address a larger market and secure more comprehensive retrofits per building.

- Provide steady resources for work on standards, R&D, technology transfer, data gathering, monitoring, and evaluation efforts.

- Establish and enforce energy efficiency performance standards for rehabilitation and construction of public and assisted housing.

- Better educate HUD area offices and housing authorities on opportunities for efficiency improvement and performance contracting.

- Set up a national insurance pool to guarantee savings from performance contracting in public and assisted housing.

- Create incentives for, and facilitate aggregated purchase of, high-efficiency appliances suitable for apartment buildings.

---

funding levels as well as poor understanding of the multifamily market's discrete subsectors and their unique barriers.

The federal government spends roughly $3 billion per year on energy bills in public and assisted housing. Yet HUD has no effective mechanism for reducing the large fraction—perhaps one-third—of these costs wasted through inefficient energy use. Efforts initiated in the early 1980s were not funded anywhere close to the authorized and necessary levels. Most of these programs have since expired and, with the decline in modernization funds, leave few resources for efficiency improvement. When modernization funding is available, energy concerns are too often neglected, resulting in lost opportunities for money-saving efficiency upgrades during other property improvement work. HUD has yet to significantly tap into the benefits possible through its performance contracting guidelines; the agency needs to do a better job of educating area offices and housing authorities about how to realize the opportunities. At the time of this writing, HUD is undergoing "reinvention" while being pressured by a change in congressional priorities away from use of federal money to meet public investment and human needs. In spite of federal reinvention and downsizing, it is important to maintain strong incentives for public housing authorities to invest in energy efficiency. Proposals to appropriate federal dollars for upgrading energy efficiency in public and assisted housing are justifiable as a sound investment for the government.

For public and assisted housing, an investment level of at least $500 million per year is warranted for energy conservation work. This is less than 15% of what is spent to subsidize energy bills and would be sufficient to reach the total federally assisted stock within 10–12 years. Yet only a small part—perhaps $10 million—of this justifiable investment level need be provided by HUD if creative ways are found to leverage private funding. A modest and specific commitment of funds, not now being made available, should be dedicated to improving the energy efficiency of the building stock for which the federal government is responsible. Taking a long-term view, there is no doubt that a decade of steady, leveraged, and targeted investments in energy efficiency upgrades would yield large net reductions in outlays to the federally assisted apartment sector.

Department of Energy

DOE needs to increase its efforts to encourage and stimulate efficiency improvement of buildings in low-income urban areas. The Clinton Administration's Climate Change Action Plan (CCAP) highlighted programs for improving energy efficiency in buildings. DOE's "Rebuild America" program seeks to foster partnerships between the federal government and cities to improve energy efficiency in existing commercial and large multifamily buildings in urban areas (DOE 1994). This new effort will be coordinated with ongoing efforts such as the DOE-HUD Initiative (Brinch 1995), DOE's Existing Building Efficiency Research Program, and Urban Consortium projects. The Environmental Protection Agency's "Energy Star Buildings" program is oriented to the commercial sector and has not been marketed to apartment building owners. However, this program could have a valuable impact by extending its reach, particularly to large, privately owned apartment buildings, which share some of the institutional characteristics of commercial business properties. In general, additional, concerted efforts are needed to ensure that the multifamily sector receives adequate attention in federal efficiency programs. It would be appropriate to require that a specified share of overall federal assistance for building efficiency be directed toward improving apartment buildings.

DOE, with support from Congress, should spend at least $10 million per year on technical programs serving multifamily housing, including efficiency standards, product research and development (R&D), commercialization, retrofit monitoring, program evaluation, and informational efforts. The department should more closely cooperate with private-sector organizations and trade associations involved in buildings and energy. DOE can do more to catalyze creative approaches to achieving efficiency investments, for instance, by inviting lender collaboration in formulating new financial instruments to facilitate investments in energy efficiency. A prototype for such approaches is Fannie Mae's financing of a demand-side contract with PG&E in California, as described in Chapter 5. DOE should make its case to Congress for such funding on the basis of the federal energy cost savings that will result as well as the broader economic and social benefits.

DOE should consider establishing an electronically networked clearinghouse for information on effective multifamily energy conservation programs, coordinating with and assisting state informational efforts. A federal role is desirable since many common technical issues exist, nationally and spanning multistate regions, and because not all states have strong programs. This activity would help cross-fertilize successful programs as well as link implementing organizations to training and technical experts within their region. An information clearinghouse should be complemented by bolstered sector characterization and evaluation data–gathering efforts. The multifamily retrofit information network can be tied together by reviving and expanding the multifamily energy conservation database (last reported by Goldman et al. 1988b, 1988c, 1988d).

Evaluations based on monitoring actual pre- and post-retrofit energy consumption should always be required for programs receiving federal assistance. DOE should systematically collect and track these results, since such scorekeeping is essential for enhancing and maintaining the effectiveness of conservation programs. It would be helpful to revise WAP funding guidelines and cycles as needed to accommodate expenses for follow-up work to secure proper building operation and persistent savings for multifamily retrofit projects. Further recommendations covering research needs that should be addressed by DOE are given at the end of this chapter.

Department of Housing and Urban Development

At HUD, the efforts to reinvent the agency must be scrutinized in light of the importance of reducing energy waste in public and assisted housing. When considering a redirection of assistance toward households rather than buildings, policy makers must account for impacts on equity and energy efficiency. The promise of affordable housing must be realistic given the state of the building stock. Some properties have energy consumption rates substantially higher than those of privately owned single-family housing. Transferring responsibility for energy bills from housing authorities to low-income tenants can create an unacceptable burden. Positive aspects of a reinvention could include creative approaches to overcome what have been roadblocks to efficiency improvement. One such roadblock is the approval burden associated with combining private financing with federal dollars for more effective efficiency investment packages. Another obstacle is the prohibition against a single contractor performing both design and implementation of retrofits, which inhibits performance contracting for assisted housing.

To rationalize HUD's energy-related efforts, the secretary should appoint an energy czar and move the Energy and Environmental program to the secretary's office. Establishing energy efficiency targets and an energy line item in the budget would be valuable mechanisms for gaining control over costs and guiding effective investments while restructuring evolves. There is authority for energy efficiency standards applicable to some property rehabilitation work, but the existing standards are out of date; an effort is underway at HUD to develop updated guidelines. Nevertheless, such standards are not now applicable to larger programs, such as Community Development Block Grant (CDBG). Therefore, consolidated programs should include provisions for energy performance standards applying to rehabilitation of existing buildings as well as to new construction.

The DOE-HUD Initiative is a promising, but as yet largely unrealized, opportunity for a coordinated federal attack on energy waste in assisted housing. With this initiative providing technical assistance and analytical support, efficiency standards plus HUD funding to leverage private investments could go a long way toward improving the energy efficiency of the nation's public and publicly assisted housing.

HUD's guaranteeing of housing authority debt for performance contracting is another useful step that can be taken. Without such a mechanism to provide security to lenders, financing for capital improvement will not be obtained, and the supply of affordable housing will erode even further. HUD might also consider creating a savings insurance pool that would lower the cost of financing efficiency improvements in public housing. The department should create additional performance contracting incentives for assisted housing—for example, by developing guidelines that would permit use of housing vouchers to assist in debt service for energy-saving retrofits.

For both public and assisted housing, HUD should actively promote performance contracting strategies to housing authorities, building own- ers, and property managers. Since utility and other private monies will largely fund the major efficiency investments that are needed, a HUD commitment on the order of $10 million per year could leverage several hundred million dollars in capital improvements. Thus, training and technical assistance regarding utility programs, performance contracting, and other innovative financing strategies is an important priority.

HUD and DOE should collaborate to create incentives for appli- ance manufacturers to offer state-of-the-art and advanced-efficiency versions of appliances and heating, ventilating, and air conditioning (HVAC) equipment suitable for apartment building applications. A valuable model for such efforts is the bulk purchase initiative recently started in New York for bringing an apartment-sized super-efficient refrigerator to market (and described under Refrigerators and Freez- ers in Chapter 3). Similar efforts should be extended to other housing authorities, and initiatives should be set up for other products, such as heat pumps, heat pump water heaters, clothes washers, and other equipment used in apartment buildings. HUD can team with the Con- sortium for Energy Efficiency (CEE), utilities, housing authorities and assisted housing trade associations, and others (such as state and local governments or community organizations) to promote purchase ag- gregation efforts for efficient appliances.

HUD should make an energetic effort to inform appraisers in- volved in the conversion, disposition, and refinancing of subsidized housing about the opportunities afforded by the Community Preser- vation Act to achieve property value enhancements through energy- efficient retrofit and rehabilitation. Such energy efficiency improve- ments add value by reducing net operating costs and thereby offer substantial opportunities for preserving affordable housing. However, opportunities are being lost because appraisers, owners, and financial entities involved in disposition and conversion are ill informed about this potential. HUD needs to engage experienced organizations in an effort to provide the information, training, and technical assistance needed to realize such property value enhancements through effi- ciency improvement. HUD should also allow sufficient owner equity return, thus providing incentives for owners to keep apartments af- fordable and to develop new affordable housing.

Other Federal Considerations

As energy policy evolves in the future, there will be a growing in- terest in the use of broad-based market incentives, such as energy or carbon taxes. Any such tax should be complemented by directing a greater portion of public resources to improving energy efficiency in buildings, particularly multifamily housing. In principle, energy price increases are helpful in providing a market signal to induce efficiency improvements. However, the same barriers to conservation that operate in the multifamily sector also obstruct an effective response to price increases alone. Therefore, both to achieve significant efficiency improvements and to offset the adverse equity consequences of higher taxation, it is imperative that pricing approaches to energy policy include provisions to devote additional resources to energy conservation programs for multifamily housing (as for low-income housing and energy conservation efforts in general).

Another long-range policy consideration is the fundamentally inequitable tax treatment of energy efficiency investments as compared with energy costs. Building owners can handle energy costs as a deductible expense. On the other hand, most retrofits, particularly the more comprehensive approaches that we find to be so valuable for apartment buildings, must be treated as depreciable capital improvements and receive less favorable tax status. Consideration must be given to reforms that would allow faster depreciation of energy efficiency retrofits and remove or limit the deductibility of energy costs. Although such tax code changes may seem radical, they are worth exploring as a way to address the historically unbalanced treatment of energy supply versus energy savings, a situation that distorts the market toward inefficient energy use.

State and Local Programs

Much program innovation occurs at the state and local levels. As summarized in Table 6-3, states should be further encouraged to use their own resources, as well as federal sources that they manage, to explore new approaches to improving multifamily building efficiency. Ways to leverage private financing hold promise. A small fuel tax increase could help fund improvements in oil-heated buildings. State public service commissions should strongly encourage utilities' efforts to strengthen and expand multifamily programs—for example, by incorporating performance contracting arrangements. Energy efficiency performance standards should be considered for all multifamily housing and retrofits that receive subsidies or other forms of financial backing from states or cities. Increased involvement is especially needed in many Sunbelt states, where relatively few successes in apartment building efficiency improvement have been reported to date. Local governments and community organizations have a crucial

Table 6-3

Recommendations for State and Local Efforts to Advance Energy Efficiency in Apartment Buildings

- Establish efficiency performance standards for housing and retrofits that receive state or local assistance or fall under local jurisdiction.

- Set up a fund for efficiency improvements in oil-heated buildings, which could be funded by a small fuel oil tax increase.

- Provide a statewide clearinghouse for information on multifamily building energy efficiency, covering sector data, technology transfer, and retrofit monitoring and evaluation results.

- Strengthen incentives for utility involvement in multifamily energy conservation.

- Use state and local funds to leverage other financing and provide state leadership to coordinate procurement of high-efficiency appliances and equipment suitable for apartment buildings.

- Make full use of available utility, federal, and state resources.

role in delivering energy conservation services to the multifamily sector. Grant programs are often provided through local governments, which also have influence in the governance of local housing authorities and assisted housing organization. This presence can provide the leverage to require that property improvements meet appropriate standards of energy efficiency.

States should coordinate with regional federal offices in collecting and maintaining multifamily sector data and retrofit performance information. States should support technology transfer working groups to make better use of regional and state resources and to streamline implementation processes. The information-gathering network should also reach out to utilities, fuel vendors, and local housing organizations, weatherization and rehabilitation implementation organizations, and researchers to enable routine gathering and sharing of energy use (billing) data for analysis, evaluation, and statewide trends tracking. The network should provide an automated degree-day reporting service for a representative set of locales throughout a state. Monitoring and evaluation must be part of all efforts. More generally, states should increase technology transfer efforts and the dissemination of successful program approaches, thereby providing a mechanism for sharing problems, solutions, and results among various implementing agencies within the state.

Table 6-4

Recommendations for Utility Programs to Advance Energy Efficiency in Apartment Buildings

- Ensure that apartment buildings receive a fair share of DSM investments.

- Take advantage of HUD performance contracting guidelines by linking DSM incentives to performance contracts and allowing cost-sharing for comprehensive retrofit packages.

- Use utility DSM funds to leverage other financing, provide low-interest loans or write-down interest rates, and provide debt servicing.

- Utilize building and HVAC trades as allies in marketing and providing comprehensive energy efficiency services.

- Include social and environmental externalities in DSM cost/benefit calculations.

Utility Programs

Utilities need to become much more active in providing electricity and gas demand-side management (DSM) conservation services to multifamily customers, who are probably not receiving a due portion of utility DSM investments. Table 6-4 lists our recommendations as to how utility programs can do a better job of reaching the apartment building sector. Efforts targeting both publicly and privately owned multifamily housing can be thought of as opportunities to "mass produce" demand-side savings, since treating such multi-unit housing can yield relatively large impacts by working with relatively few key decision makers (compared with single-family-oriented programs).

Property owners and management firms are best reached through forums familiar to them, such as publications of property-listing services for a local market. Contractors in the building and HVAC trades can be important allies; they may need training to properly address energy efficiency in apartment buildings, but once they have expertise, such firms can help a utility reach a wider market and provide more comprehensive services.

Within the multifamily sector, public housing has been hard to reach by DSM programs even though it offers substantial energy-saving opportunities. Recent changes in HUD Performance Funding System (PFS) guidelines issued in September 1991 now permit public housing authorities to retain a portion of energy bill savings generated from efficiency improvements installed under a performance contract.

Utilities should take advantage of this incentive by structuring their programs to link DSM incentives to performance contracts made with a public housing authority. In addition to providing subsidized financing, utility programs targeting multifamily public housing should consider incorporating a cost-sharing allowance, so that a public housing authority (PHA) can pay for retrofit elements that exceed the avoided cost ceiling; resident education, building staff training, and performance monitoring; and energy auditing to qualify measures based on building-specific data. Morgan (1994) provides further discussion and case studies of how utilities can more successfully reach public housing through DSM efforts.

As is recommended for public funding, utility funding should be used to leverage other forms of financing and to provide debt servicing so as to address a larger market and allow more comprehensive investments per building. In order to attract multifamily building owners, loans should carry low interest rates whenever possible.

Incorporating the value of social and environmental externalities into DSM benefit/cost calculations will help strengthen the case for energy efficiency in all electricity-using sectors. For multifamily buildings in particular, calculations should include the benefits of arrearage reduction, community economic development, and job creation, as well as all avoided environmental externalities. The value of such benefits has been estimated and applied to calculate the cost-effectiveness of a utility-sponsored weatherization program in the Northeast (Jacobson et al. 1988). Benefit/cost calculations should reflect the reduced maintenance expenses associated with energy-efficient retrofits. Aesthetics and safety benefits should be excluded from the benefit/cost test formula but included in cost-share arrangements, since they are an important factor for many building owners.

Research Needs

More information is needed on the energy consumption and building characteristics of the multifamily sector in general, as well as in particular subsectors, such as newer housing in the Sunbelt and suburban multifamily housing nationwide. This characterization will enable better estimates of program needs, financial needs, and the overall multifamily energy conservation potential.

More extensive evaluation research is needed for the sector, for which a relative lack of information about energy conservation program effectiveness remains an issue. Evaluation research must be organized to reflect the diversity of the multifamily housing stock, providing separate coverage according to geographic region, building type, and system type. The variety of institutional and program delivery situations (public, publicly assisted, or private; small or large buildings; utility, state/local, or community based) must also be considered. Evaluation methodologies appropriate for each institution and region should be developed and incorporated as an inherent part of future programs.

DOE's ongoing Weatherization Assistance Program (WAP) review for the multifamily sector was scaled down from a full survey to case studies of five programs (in Chicago; New York; Seattle; Springfield, Massachusetts; and St. Paul). Although these case studies provide valuable snapshots of the sector, a more detailed and representative look at what works and what is cost-effective is needed. Ongoing energy-consumption data gathering and evaluation analyses for multifamily WAP efforts should be built into future funding cycles. Monitoring and evaluation should be required for government-funded retrofit programs by all agencies. Finally, DOE should provide an information clearinghouse service—linked to national laboratory, state-based, and other research programs—for multifamily retrofit evaluation results, organized to account for the geographic, physical, and institutional diversity of the multifamily stock.

The fragmentation and small-business basis of the building technical services sector has long provided a compelling rationale for a strong federal R&D role. The need is particularly urgent for multifamily buildings because of their greater complexity and the woeful lack of private expertise that has been reported by conservation practitioners. Suggestions include expanding national laboratory-based programs to better address the multifamily sector and providing federal assistance for the R&D efforts of state and local conservation organizations. Research sponsored by utilities and utility organizations has helped in the multifamily arena. These efforts should be expanded. Concerted federal and state research efforts are needed for oil-heated buildings, in which there is not a utility interest. Good coordination of government and utility research programs will help guarantee appropriate targeting and timely transfer of research results to practitioners in the field.

Technical topics needing further research in the multifamily sector include mechanical systems; ventilation, infiltration, internal gains, and building mass dynamics (especially the interactions among various components of a multi-unit structure); moisture and air quality; occupant comfort and behavioral factors; building shell heat loss; and energy use by appliances and ancillary equipment (elevators and so forth). There is a need for refinement and validation of computer-based tools for auditing multifamily buildings and analyzing the performance of retrofits, as well as greater efforts in dissemination and training for the use of these tools.

As discussed in Chapter 3, techniques for heating cost allocation provide ways to estimate energy consumption in an individual apartment so that the tenants can be billed in proportion to their energy use. Heating cost allocation is common in Europe, but a variety of obstacles have kept it from widespread use in the United States. One barrier is a lack of standardized guidelines for implementing heating cost allocation in multifamily housing. Since many of the issues are generic and cut across state lines, this is a worthy area for federal attention. Standards for heating cost allocation should cover building efficiency, equipment and installation, equitable allocation of unmetered consumption, meter reading and billing procedures, disclosure, dispute resolution, and tenant education. The European experience offers models on which to draw; Hewett (1988a) has researched the issue and provides more specific recommendations.

The importance of treating the building as a system, including all of the human actors involved, is a key theme identified in our review. Behavioral research has contributed to our understanding of how to achieve efficiency improvement in all building sectors. Further research along these lines is crucial for the multifamily sector because of the diverse interests of the key actors: owners and management, building staff, and residents. Topics include the comfort factors and the interaction of people with control systems, such as thermostats and energy management systems; implications for equipment designs; and ways to improve and sustain resident education. Social science research should also address building staff and their decision making, a topic that is particularly important in this sector but which has received inadequate attention.

Coincident with the decline of federal energy efficiency R&D throughout the 1980s, states came to recognize the importance of energy-related R&D to their economies. By 1992, eight states were spending $39 million per year—equivalent to one-fifth of DOE's total conservation and renewable energy budget—on energy efficiency R&D programs (Harris et al. 1992). New York State in particular has devoted considerable effort to energy use research and conservation program development for multifamily buildings. Such state efforts should be supported and expanded, and DOE can assist through coordination and cofunding.

A collaborative approach to federal R&D is being taken in many areas of technology. For example, the federal government can establish cooperative R&D arrangements with energy supply and energy efficiency industries. Such arrangements, with perhaps a larger federal share, should be further explored for the "appropriate technology" being pursued by multifamily energy conservation researchers at state and local levels. Many of the technical people in state, municipal, nonprofit, and utility programs have the skill and desire for research as an adjunct to their retrofit work but rarely have the time and money to do much of it; a large part of what does get done is through their individual dedication. Relatively few federal dollars (compared with what is being spent in other sectors) could go a long way toward delivering valuable research results. Cooperative R&D efforts would also facilitate "technology transfer," provide ready feedback from field trials to federal researchers, and help expand the pool of energy efficiency researchers and practitioners nationwide.

Making Efficiency the Norm

Throughout our discussion, we have pointed out the value of linking energy efficiency performance criteria to various programs, either as performance and equipment standards applicable during rehabilitation efforts or as a contingency on financing. Over the long run, a more aggressive stance on buildings performance will be necessary to bring the U.S. apartment building stock up to cost-effective levels of energy efficiency. Energy-efficient building codes and equipment standards are necessary to ensure that new construction does not result in additions to the stock of buildings that must be subsequently retrofit to eliminate energy waste. For existing buildings, incentive programs, voluntary standards, utility DSM programs, and technical assistance can make a meaningful dent in multifamily housing energy consumption over the coming years. However, this progress is likely to be localized in a few regions and subsectors of the overall market. Absent standards, energy waste will continue elsewhere, with efficiency investments still inhibited by the many market barriers that exist in this sector.

Energy efficiency standards can be applied to existing buildings at time of sale (the "retrofit ordinance" approach). Although most building owners are likely to oppose efficiency standards as costly or unnecessary regulation, it will be worth making the case for flexible, performance-based standards that can be coupled with incentives and would help enhance property values as well as the availability of affordable housing. Standards for existing buildings would need to be packaged to provide adequate lead time, attractive financial incentives, and technical assistance, and involve coordination with private-sector partners including equipment manufacturers, utilities, buildings services trades, and energy service companies in whom building owners and management firms can develop trust. Initially, building efficiency standards can be best advanced voluntarily by states, taking advantage of federalism at its best. An alliance of progressive interests—state and local governments, low-income and affordable housing advocates, renter organizations, energy conservation organizations, utilities, and energy efficiency vendors—should work to establish standards and induce investments that can make energy efficiency the norm in apartment buildings throughout the United States.

Acronyms

ACEEE	American Council for an Energy-Efficient Economy
AFUE	Annual Fuel Utilization Efficiency
AHS	American Housing Survey
ANL	Argonne National Laboratory
ASHRAE	American Society of Heating, Refrigerating and Air-Conditioning Engineers
ASTM	American Society for Testing and Materials
BCR	benefit/cost ratio
BECA	Building Energy Use Compilation and Analysis
BECo	Boston Edison Company
CABO	Council of American Building Officials
CACS	Commercial and Apartment Conservation Service
CCAP	Climate Change Action Plan
CCC	Citizens Conservation Corporation (Boston)
CCS	Citizens Conservation Services (subsidiary of EUA Cogenex Corp., Lowell, Massachusetts)
CDBG	Community Development Block Grant
CEE	Center for Energy and Environment (Minneapolis), formerly Center for Energy and Urban Environment, and earlier the Minneapolis Energy Office (MEO)
CEE	Consortium for Energy Efficiency (Boston)
CESF	Chicago Energy Savers Fund

CFL	compact fluorescent lamp
CHA	Chicago Housing Authority
CHAS	Comprehensive Housing Affordability Strategy
CIAP	Comprehensive Improvement Assistance Program
CIRA	Computerized Instrumented Residential Audit
CL&P	Connecticut Light and Power Company
CNT	Center for Neighborhood Technology (Chicago)
CRA	Community Reinvestment Act (1977)
CSA	Community Services Administration
DHW	domestic hot water
DOE	U.S. Department of Energy
DSM	demand-side management
EBER	Existing Building Efficiency Research (program of U.S. DOE)
ECM	energy conservation measure
ECPA	Energy Conservation and Production Act (1976)
EERC	Environment and Energy Resource Center (St. Paul), formerly Energy Resource Center (ERC)
EIA	Energy Information Administration
EMS	energy management system
EPA	Environmental Protection Agency
EPACT	Energy Policy Act (1992)
EPCA	Energy Policy and Conservation Act (1975)
EPRI	Electric Power Research Institute (Palo Alto, California)
ERC	Energy Resource Center (St. Paul)
ERTA	Economic Recovery Tax Act (1981)
ESCO	energy service company
ESTSC	Energy Science and Technology Software Center (Oak Ridge, Tennessee)
EUA	Eastern Utilities Associates (Massachusetts)
EURWAP	Expiring Use Restriction Weatherization Assistance Program (Massachusetts)
Fannie Mae	Federal National Mortgage Association (former name)
FHA	Federal Housing Administration
FmHA	Farmers Home Administration

Freddie Mac	Federal Home Loan Mortgage Corporation
HCD	Housing and Community Development Act
HERS	home energy rating system
HFA	Housing Finance Agency
HHS	U.S. Department of Health and Human Services
HUD	U.S. Department of Housing and Urban Development
HVAC	heating, ventilation, and air conditioning
IREM	Institute for Real Estate Management
IRP	integrated resource planning
IRS	Internal Revenue Service
LBL	Lawrence Berkeley Laboratory
LCD	liquid crystal diode
LCP	least-cost planning
LIHEAP	Low Income Home Energy Assistance Program
LIHPRHA	Low Income Housing Preservation and Resident Homeownership Act (1990)
MCS	Model Conservation Standards
MEO	Minneapolis Energy Office (now CEE, Minneapolis)
MHFA	Massachusetts Housing Finance Agency
MPS	Minimum Property Standards
NAECA	National Appliance Energy Conservation Act (1987)
NBS	National Bureau of Standards (now NIST)
NCAT	National Center for Appropriate Technology
NCLC	National Consumer Law Center
NFRC	National Fenestration Rating Council
NIST	National Institute of Standards and Technology
NYCHA	New York City Housing Authority
NYPA	New York Power Authority
NYSERDA	New York State Energy Research and Development Authority
O&M	operations and maintenance
OPEC	Organization of Petroleum Exporting Countries
ORNL	Oak Ridge National Laboratory
OTA	Office of Technology Assessment
PFS	Performance Funding System

PG&E	Pacific Gas and Electric Company (California)
PHA	public housing authority
PILIRR	Partnerships in Low-Income Residential Retrofit
PRISM	Princeton Scorekeeping Method
PSC	Public Service Commission
PVEA	Petroleum Violation Escrow Account
RAP	Ramsey Action Program (Minnesota)
RCS	Residential Conservation Service
R&D	research and development
RECS	Residential Energy Consumption Survey
REEP	Residential Energy Efficiency Program
RRR	Residential Rental Retrofit (program for buildings standards in Minnesota)
SCL	Seattle City Light
SEECB	Solar Energy and Energy Conservation Bank
SEER	seasonal energy efficiency rating
SERP	Super-Efficient Refrigerator Program
SIR	savings-to-investment ratio
SyrESCO	Syracuse Energy Services Company
UCETF	Urban Consortium Energy Task Force
UDAG	Urban Development Action Grant
VEIC	Vermont Energy Investment Corporation (Burlington)
VHFA	Vermont Housing Finance Agency
WAP	Weatherization Assistance Program
WECC	Wisconsin Energy Conservation Corporation
WHEEL	Wisconsin Heating Energy Efficiency Label
WPSC	Wisconsin Public Service Commission

Resource Organizations for Energy Efficiency in Apartment Buildings

A variety of organizations have been involved in aspects of apartment building energy conservation over the years. The list compiled here includes those that have recently reported experience, program activities, or information provision services that would be helpful for others interested in pursuing energy efficiency improvements in this sector. Organizations are identified according to the types of services provided and regional scope of their activities (although some organizations listed for a given region may also be able to provide services nationwide). Letter codes in the listing are defined as follows:

Services
- C = Consulting
- I = Information only
- P = Programs and implementation
- R = Research

Regions
- E = Eastern, Northeast
- N = Northern Plains, Midwest
- S = Sunbelt (Southeast, South Central)
- W = Western
- U = U.S. Nationwide

Table B-1 at the end of the appendix provides a cross-reference to the numbered organization list by region.

221

Name (alphabetical listing)	Types of Services	Regional Scope
1. Alliance to Save Energy 1725 K Street NW, Suite 509 Washington, D.C. 20006 202-857-0666	I	U
2. American Council for an Energy-Efficient Economy 1001 Connecticut Avenue, N.W., Suite 801 Washington, D.C. 20036 202-429-8873	I	U
3. Ann Arbor, City of Community Development Department 100 North Fifth Avenue PO Box 8647 Ann Arbor, MI 48107 313-994-2912	P	N
4. Argonne National Laboratory 9700 South Cass Avenue Argonne, IL 60439 708-252-8688	R	U
5. Austin, City of Environmental and Conservation Services 206 East Ninth Street, 2 Commodore Plaza Austin, TX 78701 512-499-2818	P	S
6. Barakat and Chamberlin 1800 Harrison Street, 18th Floor Oakland, CA 94612 510-893-7800	C	W
7. Bonneville Power Administration PO Box 3621 Portland, OR 97208 503-230-3000	P	W
8. Boston Edison Company 800 Boylston Street Boston, MA 02199 617-424-2000	P	W

Name (alphabetical listing)	Types of Services	Regional Scope
9. California Energy Commission 1516 Ninth Street, MS-22 Sacramento, CA 95814 916-654-4000	I	W
10. Center for Energy and the Environment 100 North Sixth Street, Suite 412A Minneapolis, MN 55403-1520 612-348-6829	P,R	N
11. Center for Neighborhood Technology 2125 West North Avenue Chicago, IL 60647 312-278-4800	P	N
12. Citizens Conservation Corporation 530 Atlantic Avenue Boston, MA 02210 617-423-7900	P,R	U
13. Citizens Conservation Services EUA/Cogenex Corporation 100 Foot of John Street Lowell, MA 01852-1197 508-656-3502	C	U
14. Conservation Services Group 441 Stuart Street Boston, MA 02116 617-236-1500	C	E
15. Conserve, Inc. 505 Eighth Avenue, Suite 1805 New York, NY 10018 212-564-5353	P	E
16. Consortium for Energy Efficiency 1 State Street, Suite 1400 Boston, MA 02109 617-589-3949	I	U
17. Domus Plus 408 North Grove Oak Park, IL 60302 708-386-0345	C,R	N

Name (alphabetical listing)	Types of Services	Regional Scope
18. Ecotope, Inc. 2812 East Madison Seattle, WA 98112 206-322-3753	C,R	U
19. Electric Power Research Institute PO Box 10412 3412 Hillview Avenue Palo Alto, CA 94303 415-855-2000	R	U
20. EME Group 135 Fifth Avenue New York, NY 10010 212-529-5969	C,R	E
21. Energy Conservatory 5158 Bloomington Avenue South Minneapolis, MN 55417 612-827-1117	C,R	U
22. Energy and Environment Resource Center 427 St. Clair Avenue St. Paul, MN 55102 612-227-7847	C,R	N
23. Energy Management & Research Associates 49 Murdock Court, Suite 6J Brooklyn, NY 11223-6414 718-332-2926	C	E
24. Florida Solar Energy Center 1679 Clear Lake Road Cocoa, FL 32922-5703 407-638-1000	R	S
25. GRASP 3500 Lancaster Avenue Philadelphia, PA 19104 215-222-0318	P,R	E

Name (alphabetical listing)	Types of Services	Regional Scope
26. *Home Energy* Magazine 2124 Kittredge Street, No. 95 Berkeley, CA 94704 510-524-5405	I	U
27. Illinois, State of Department of Energy and Natural Resources Springfield, IL 62701 217-785-2373	P	N
28. Lawrence Berkeley Laboratory Center for Building Technology Building 90-3000 Berkeley, CA 94720 510-486-4834	R	U
29. Michigan Public Service Commission 6545 Mercantile Way Lansing, MI 48909 517-334-6445	I	N
30. Multifamily Loans Program Connecticut Housing Investment Fund 121 Tremont Street Hartford, CT 06105 860-233-5165	P	E
31. National Center for Appropriate Technology PO Box 3838 Butte, MT 59702 406-494-4572	I	U
32. National Consumer Law Center 18 Tremont Street, Suite 400 Boston, MA 02108 617-523-8010	I	U
33. National Renewable Energy Laboratory 1617 Cole Boulevard Golden, CO 80403 303-275-3000	I	U

Name (alphabetical listing)	Types of Services	Regional Scope
34. New England Power Service Corporation 25 Research Drive Westborough, MA 01582 508-355-9011	P	E
35. New York City Weatherization Coalition 505 East Eighth Avenue, 18th Floor New York, NY 10018 212-729-3902	P	E
36. New York State Energy Research & Development Authority 2 Rockefeller Plaza, 14th Floor Albany, NY 12223 518-465-6251	R	E
37. North Carolina Alternative Energy Corp. PO Box 12699 Research Triangle Park, NC 27709 919-361-8000	R	S
38. Oak Ridge National Laboratory Energy Division PO Box 2008 Oak Ridge, TN 36830 615-574-5187	P	E
39. Pacific Energy Associates 510 North Tomahawk Portland, OR 97217 503-626-8096	C	W
40. Pacific Gas & Electric Company 3400 Crow Canyon Road San Ramon, CA 94583 510-866-5449	P	W
41. Portland Energy Office 1211 SW Fifth Avenue, Room 1170 Portland, OR 97204-3711 503-823-7222	P	W

Name (alphabetical listing)	Types of Services	Regional Scope
42. Proctor Engineering Group 818 Fifth Avenue, Suite 208 San Rafael, CA 94901 415-455-5700	C,R	U
43. Public Service Company of Oklahoma Good Cents Energy-Efficient Apartments Program PO Box 201 Tulsa, OK 74102-0201 918-599-2000	P	S
44. Queens College Department of Urban Studies Flushing, NY 11367 718-997-5134	R	E
45. Ramsey Action Program 3315 Labore Road Vadnais Heights, MN 55110 612-482-8260	P	N
46. Residential Energy Services 320 Huron Street Fairchild, WI 54741 715-334-2271	C,R	U
47. Sacramento Municipal Utility District 6201 S Street Sacramento, CA 95817 916-732-5494	P	W
48. St. Louis Energy Management Program 411 North Tenth Street St. Louis, MO 85101 314-464-7071	P	S
49. Seattle City Light 1015 Third Avenue Seattle, WA 98104 206-625-3000	P	W

Name (alphabetical listing)	Types of Services	Regional Scope
50. Sun Power Consumer Association 5160 Parfet Street, A-3 Wheat Ridge, CO 80033 303-467-0521	C	W
51. Synertech Systems Corporation 472 South Salina Street, Suite 410 Syracuse, NY 13202 315-422-3828	C,R	U
52. SyrESCO 614 S. Salina Street Syracuse, NY 13202 315-451-3305	C,R	E
53. Tacoma Public Utilities PO Box 11007 Tacoma, WA 98411 206-383-9600	P	W
54. Underground Space Center 790 C.M.E. Building University of Minnesota Minneapolis, MN 55455 612-624-0066	R	N
55. U.S. Department of Energy Office of Buildings Technology 1000 Independence Avenue SW Washington, DC 20585 202-586-9445	I	U
56. U.S. Department of Housing and Urban Development Office of Community Planning and Development 451 Seventh Street SW Washington, DC 20410-7000 202-708-3363	I	U

Name (alphabetical listing)	Types of Services	Regional Scope
57. Vermont Energy Investment Corporation 7 Lawson Lane Burlington, VT 05401 802-658-6060	P	E
58. Wisconsin Energy Conservation Corporation 3120 International Lane Madison, WI 53704 608-249-9322	C,R	N

Table B-1

Cross-Reference to Organization List by Region

Eastern (Mid-Atlantic and Northeast)	14, 15, 20, 23, 25, 30, 34, 35, 36, 38, 44, 52, 57
Northern (Northern Plains, Midwest)	3, 10, 11, 17, 22, 27, 29, 45, 54, 58
Sunbelt (Southeast, South Central)	5, 24, 37, 43, 48
Western	6, 7, 8, 9, 39, 40, 41, 47, 49, 50, 53
U.S. (nationwide)	1, 2, 4, 11, 12, 16, 18, 19, 21, 26, 28, 31, 32, 33, 42, 46, 51, 55, 56

Audit Tools for Apartment Buildings

Chapter 3 discussed various approaches to energy audits for apartment buildings, stressing the need for a fact-finding-based approach that considers all aspects of the "building as a system," considering the structure, energy-using equipment, and human actors, including residents, staff, and management. In some conservation efforts, resource constraints or a narrow program scope may prevent an ideal assessment of every building. But even in the most limited programs, learning about the operating environment of the buildings to be addressed is crucial.

This appendix describes some of the various diagnostic tools for helping auditors understand energy use in an apartment building. Analysis of actual consumption records is always fundamental. We also highlight two technical areas, combustion system testing and building air flow diagnosis, and provide an overview of integrating approaches, including audit handbooks and computerized audit programs.

Not all tools are used in all circumstances—for example, in limited, prescriptive programs, a checklist of eligible measures may be the only "tool." Generally, however, substantial information gathering is needed. A building's energy use history is an important place to start. Audit tools can be thought of as ways to help the auditor understand why the building does not work as it should:

- If the roof or other part of the building shell fails to keep out rain, visual inspection or a moisture meter (or sometimes a thermographic camera in inaccessible flat roofs) helps identify the problem and shows where the repair is required.

- If the building skin is intended to isolate external and interior air, a blower door or tracer gas tests can show how well it is succeeding and where it is failing.

- If installed insulation performs poorly, a thermographic camera reveals where it is failing to adequately block heat transfer.

Of course, auditors must not overlook the obvious. Questioning tenants or taking spot temperature measurements can identify apartments with comfort problems and guide more detailed physical inspections. Moreover, an audit must place energy-related information in the context of the broader set of concerns related to building operation, including health and safety measures as well as general repair work. As Len Rodberg points out:

> Repair measures have to be done, even if they don't save energy. However, they are necessary for doing measures that do save energy, and it is important to describe them and track their costs. There are lots of multifamily weatherization measures like this, akin to repairing roof leaks so that insulation may be installed. [Rodberg 1994]

The complexities involved in analyzing apartment buildings should make apparent the value of a set of tools that can measure what is going on in a building as well as organize the myriad pieces of information that must be tracked and analyzed.

Consumption Data Analysis

A basic element of energy analysis in apartment buildings is obtaining utility bills, either from the building's management or from the utility. High energy use is a valuable predictor of savings potential and therefore of whether and where successful efficiency investments can be made. Documenting pre-retrofit (base) energy use and costs, understanding the metering configuration (master versus unit-level), and assessing utility billing procedures and rate structures are essential steps. A building's consumption history forms the foundation for projecting retrofit impacts on energy costs, post-retrofit cash flow, and evaluation of achieved savings. (Techniques for billing data analysis are discussed under Billing/Statistical Analysis in Chapter 4.)

Annual totals are much less useful than monthly statements that allow extraction of base load consumption and assist in identifying particular anomalies that may have occurred at some specific time. Accurate records of meter reading dates are needed to compute precise Btu-per-degree-day indices using local weather data. Collecting monthly energy consumption data can be a tedious task.

This difficulty is undoubtedly why so few programs track savings or target their investments to pre-retrofit consumption. Obtaining accurate consumption data is particularly challenging for oil-heated buildings, common in New York and other parts of the Northeast. In this case, an iterative approach, using a consumption data analysis method such as PRISM (Princeton Scorekeeping Method), can be used to detect and correct for outliers and obtain meaningful baseline consumption estimates (Fels and Reynolds 1993).

The state of Montana, working with utilities, has developed a computer program for automatically retrieving billing data on all clients of their energy assistance program (Poole 1989). Each month, the utilities download clients' files in a format compatible with both the energy audit and program progress-tracking file. The main obstacles to implementing similar programs elsewhere are (1) the need to get fuel release forms from all tenants in individually metered buildings and the complexity of totaling and matching files; (2) the inconsistency and archaic nature of many utility billing records; and (3) the irregularities of delivery and partial tank fillings for oil and other nonmetered fuels.

For more detailed analysis, submetering can be a useful way to obtain data for individual apartments in a master-metered building or for particular appliances and end uses. The major challenge is the need to hard-wire a recording device between the main meter and the end-use load to be monitored. This effort is not too daunting for electricity, but it is expensive and intrusive for gas or oil service. For plug-in appliances, secondhand kilowatt-hour meters (as utilities retire them for new digital units) can serve as inexpensive diagnostic tools. Sophisticated electronics have also greatly improved this technology. Various devices can be connected to electrical loads to monitor consumption (some have been used to provide feedback to occupants) and also act as diagnostic tools.

For fuel-fired equipment, run-time metering can track the number of hours a particular device (for example, a furnace or water heater) is running. Multiplying the run time by the firing rate yields the energy consumption over the monitoring period. This approach is often the only inexpensive way to "submeter" nonelectric fuel use. Various test probes and sensors have been developed to measure a number of variables useful in auditing.

Combining submeter or run-time information with other data, such as temperature measurements, can provide a valuable picture of equipment and system performance. Solid-state data loggers are now relatively inexpensive; eight-channel data loggers with an array of sensors to record temperatures, kilowatt-hours consumption, run times, events, and many other variables are also available at less than $1,000 each.

Combustion Equipment Diagnostics

Combustion efficiency testing is a fundamental step in any apartment building having oil- or gas-fired equipment. Steady-state testing provides relatively quick information about a significant portion of energy use. For most smaller boilers and furnaces, standard procedures are available for estimating the Annual Fuel Utilization Efficiency (AFUE) or seasonal efficiencies from these tests and other observations. Unfortunately, for boilers larger than 300,000 Btu/hr, as found in many larger apartment buildings, there is no agreed-upon mechanism for determining the AFUE. As noted in Chapter 3, estimating boiler seasonal efficiency is important for such systems, but estimation techniques have not yet evolved to the level of a widely accessible audit tool. Besides basic efficiency testing, most auditors also usually check for draft; carbon monoxide generation; gas or oil leaks; leaks in the combustion chamber and heat exchanger; clogged fire tubes; poor flame dispersion in the firebox; malfunctioning or unsafely wired controls, pumps, fans, and damper motors; temperature and pressure control settings; and adequate combustion air. For buildings with forced-hot-air systems or other air-handling equipment, it has become standard to check for furnace room depressurization and backdrafting while the air handler is operating.

Review of past maintenance records provides insights as to the past and present condition of the heating plant as well as annual maintenance cost information valuable for determining the cost-effectiveness of system replacement or major upgrading. Distribution system diagnostics are typically required before, during, and after any major retrofit process. Thermal imaging of steam traps, tracking operation of air vents, and monitoring of delivered temperatures throughout the building can help in establishing a well-balanced steam system.

Air Leakage Measurement

Measurement of air leakage in large apartment buildings is usually more complicated than in smaller single-family structures. Among the available techniques, some are still in the category of research methods rather than routine audit tools. Practically speaking, air leakage identification and measurement techniques can be used in a "house-doctoring" approach to find and fix infiltration and ventilation problems in apartments and throughout a building. Experience and available data were reviewed under Ventilation and Air Leakage in Chapter 3. Among the most sophisticated methods are various tracer-gas techniques developed for multizone buildings, as applied by Francisco and Palmiter (1994), for example, and generally described by Sherman et al. (1989). Although out of reach for many auditors, such methods are the most reliable available and are worth considering in programs in which many similar buildings are to be addressed.

A more accessible technique is use of blower doors, which can help diagnose air flow paths in individual apartments as well as quantify the magnitude of contribution to air leakage of individual components or systems. Obviously, blower door tests cannot be done on every apartment in a large building, but identification of "typical" air flows in a few representative apartments can help establish a sound overall retrofit strategy.

Identifying the contribution of window air leakage versus that of other components is important for estimating the savings potential of replacement units or other modifications. A single blower door can be used to determine the leakage in a typical apartment by running controlled pressurization tests with and without the windows blocked off—for example, using plastic film and tape.

By running such tests in one or two apartments in a multiunit building, an auditor can gain insights into how the other similarly constructed units perform. In some buildings, more insight can be gained by running multiple blower doors in several adjoining units at the same time. However, the complexity and expense of such multizone tests still place them more in the category of a research tool than a routine audit procedure. NYSERDA (1995) discusses results, limitations, and promising procedures for air leakage measurement in apartment buildings using one or two blower doors. The standard test method for operating blower doors is ASTM (1988).

In high-rise structures, thorough diagnostics of actual air flow patterns between apartments and common spaces can be valuable. Measurements would be needed at high, low, and intermediate elevations and under varying climate and mechanical systems operations conditions. Such analysis can be made with the assistance of smoke sticks and accurate air pressure measuring gauges (such as digital manometers) to measure the direction and magnitude of air pressures under varying conditions. Research is presently underway at Lawrence Berkeley Laboratory and other institutions to develop methodologies for estimating the rate of flow between these various spaces using advanced pressure diagnostics.

Infrared thermography can help identify moisture problems, convective bypasses, and large convective flows (that is, points where air is infiltrating and exfiltrating). Thermography or less expensive temperature probes, in conjunction with individual apartment blower door measurements, can distinguish between air leakage from a colder (or warmer) exterior and leakage from the rest of the building. Thermography is particularly useful for flat-roof buildings where the integrity of the roof may be in question.

In buildings with a central air-handling unit (usually only found in "upscale" apartments with central air conditioning), it is possible to intentionally unbalance the air handler to act as a giant blower door and, with smoke sticks or infrared scanners, go about identifying air leakage sites (Wallace 1985). It may also be possible to make a rough estimate of overall air leakage on the basis of measured fan amperage and the unit's specifications by running pressurization or depressurization tests. Researchers have also used tracer-gas techniques in large buildings with central air handlers (ASTM 1983; Harrje 1984; Commoner and Rodberg 1986). This technique is likely to be a luxury that few programs can use on a regular basis. For a large, well-controlled project, it could be worthwhile to contact researchers at Lawrence Berkeley Laboratory to enlist their expertise with tracer-gas measurements and other advanced methods. Diagnosing infiltration and ventilation problems in apartment buildings is clearly a subject on which much more research is needed to develop practical tools for widespread use.

Audit Handbooks and Workbooks

Historically, the checklist approach to energy audits gave rise to a variety of multifamily energy conservation handbooks and audit workbooks. Some of these materials remain useful as introductions to what can be encountered in the multifamily sector or as part of a more comprehensive audit approach as discussed earlier. Workbooks and paper audits fall essentially into three categories: (1) intake forms that document the present condition of existing systems and building elements; (2) guidebooks that instruct the auditor as to what elements to look for in the multifamily building and how to look at them; and (3) calculation sheets that estimate the savings from individual measures. Such handbooks are precursors of, and have been largely superseded by, computerized energy audit tools. They typify the engineering approach to retrofit planning frequently taken in early efforts. Although such procedures are poor predictors of energy savings, the compiled information can provide useful orientation for more holistic and measurement-based approaches.

An example of the first category is the set of multifamily audit documentation guidelines developed by the state of Illinois (Illinois DCCA 1992, 1994). Detailed sheets are used to characterize the condition of heating systems typical in the region, including identification of steady-state efficiency, carbon monoxide, and other safety issues. However, the sheets do not mention the heating load or even building size and other pertinent characteristics. There is a sheet for listing recommended retrofits, but no indication of either the savings potential or how these selections were derived. In Chicago, mechanical retrofit recommendations are made by designated contractors who work with the city.

A good example of an auditor's guidebook is the HUD *Section 8 Housing Inspection Manual* (Abt Associates 1989). Designed to meet periodic field inspections of designated properties, this guide focuses mostly on structural and safety concerns. It contains little material dealing with mechanical systems, and again, the issues covered are mostly oriented to safety rather than to energy efficiency. A companion document is the HUD *Energy Conservation for Housing* workbook (Perkins & Will and the Ehrenkrantz Group 1982), covering 50 retrofit options in four categories: architectural, heating systems, secondary systems (mostly domestic hot-water and cooling), and electrical systems. Each option has a one- or two-page description, discussion of concerns and applicability, and an illustration of the device. A cost/benefit calculation for each measure is based on local heating degree-day zones and fuel costs. Measures are not prioritized, and interactive effects are not considered, other than pointing out some measures that are mutually exclusive. A similar approach is taken in the New York State *Multifamily Housing Energy Conservation Workbook* (the "Gray Book," as named for its cover; New York State Energy Office, undated).

The Minnesota Department of Public Service developed a *Multifamily Building Energy Audit, Technical Manual* and accompanying training program with an extensive resource packet (MDPS 1987). Designed for weatherization auditors and inspectors for utility or Farmers Home Administration (FmHA) programs, the procedure is known affectionately as the "Maxi Audit." It is a step beyond the checklist approach in that it seeks to provide decision makers (auditors) with the knowledge and tools to make informed decisions rather than imposing a restrictive list of retrofit opportunities. The massive document represents diverse expertise, but the focus is clearly on building a firm understanding of the first principles and the wide range of available options. The audit itself includes 5 pages of detailed information characterizing the present structural and mechanical systems. To this is attached a 2-page summary of recommended measures, including estimated savings, costs, and paybacks. The balance of the paperwork consists of 45 single-page fact sheets that explain the function and rationale behind each recommended retrofit and summarize the savings opportunities for each. These sheets are designed to educate the building owner. The calculations relating to each sheet are included in the remaining 144 pages of formulae, charts, and tables. The Maxi Audit is not a concise package. Substantial effort is required for an auditor to climb the learning curve and master the material. The end result, however, is an understanding not only of how to calculate savings estimates, but also of the basis for the calculations.

Computerized Audit Software

Much of the painstaking work associated with a detailed building audit has been simplified—at least, somewhat—with the advent of computerized energy audit tools. Nevertheless, energy audit software should be thought of as a complement to, rather than a replacement for, the experience of an expert practitioner. As with any tool, its value depends on the skill of the user. Calibrating the parameters of a computer audit to a building's actual consumption data is a critical step. When discrepancies are found, the auditor should reexamine the input data and modeling assumptions to identify what might be amiss or what else might be going on in the building to explain the discrepancies. The revised data should be entered into the program and the process repeated until there is a reasonable fit between the computer model and actual consumption.

The project manager should then reevaluate the building in terms of what retrofits are actually possible. Can insulation be added? Windows replaced? Burner and lighting retrofits made? Control and hot-water modifications devised? Questions about a host of other details surrounding heating systems controls and conditions of the delivery system must be answered; a computerized audit tool should help lead one systematically through these questions and organize the answers. To accurately evaluate retrofit measures, detailed measurements of a variety of system components are needed, as well as information enabling reasonable estimates of how the measurements would change after retrofit. This level of knowledge cannot be expected from computer software itself. Therefore, experience and understanding, as reflected in the technical expertise of the auditors and contractors, are needed to complement the computer audit.

Audit software can help perform tedious cost/benefit calculations. The savings from proposed measures can be evaluated and compared with up-front costs so that measures can be prioritized. Interactions among measures should be built into the algorithms, and the estimated savings from each measure on a list of retrofit options should be adjusted under the assumption that previous measures are implemented first. Given these adjustments, the software should readjust the priority list and repeat the evaluation process until a maximum cost/benefit ratio is obtained. The audit should be able to provide a comparison of the various scenarios, giving both estimated energy savings and the relative cost-effectiveness of the proposed packages, based on the fiscal requirements of the funding parties. Run times should be short enough that the auditor can easily change input data or select various combinations of measures and be able to get comparative analyses without having to wait overnight for the calculations to be made.

Retrofit planners do not typically rely on computerized audits as the primary decision tool. Rather, the planning process is one of careful analysis of building load, detailed system diagnostics, and then strategizing as to the most likely retrofit candidates, primarily from past experience and the condition of the building as found. The computer programs are most often used primarily for purposes of confirmation of measured building load and as justification for or against known assumptions regarding the building. For example, an experienced auditor does not need to run a computer simulation to know that wholesale window change-outs are not likely to be a cost-effective retrofit in a building with a poorly controlled heating system. The auditor may know from experience that an outdoor reset controller can predictably shave a certain percentage from hydronic heating energy use in a given locale. The building owner or the financial institution, on the other hand, may need to see more detailed calculations to be convinced.

"Four years of computer-assisted audits have made me a better auditor," reflects Andy Padian of the New York City Weatherization Coalition (Padian 1994a). Padian uses the CIRA-derived EA-QUIP program (see below) to aid in audit and retrofit planning for apartment buildings. He notes how on a few occasions, the program has told him that "my building diagnosis was wrong, and 'it' was right." Use of audit software has helped him learn which changes in a building have the largest impacts on fuel use. Padian also notes how the program helps make the case for what to do as well as what not to do: "To my complete satisfaction, window replacements show a virtually insignificant change in fuel usage, even when factoring in the combined effect of increased R-value and decreasing infiltration. Owners typically want window replacements and we typically don't want to pay for them."

In a follow-up interview, Padian says, "The EA-QUIP audit picks up a small percentage of the problems we don't see, but an audit can't predict the need to downsize a boiler . . . it sometimes tells us that some things that look good just aren't cost-effective. On the other hand, a boiler may be a piece of junk and have lots of safety problems, but is running at 75% efficiency. The audit doesn't necessarily override our decision" (Padian 1994c). To date, the available software does not incorporate repair and maintenance costs into cost/benefit calculations, but such enhancements are planned. Padian concludes that "EA-QUIP has improved our effectiveness in dealing with larger and more sophisticated building owners, and it has supported many agencies in getting close to dollar-for-dollar matching funds from owners of rental properties."

A good audit program provides an organized framework to account for all the variables determining energy use. It should incorporate data gathered from diagnostic procedures, such as furnace or boiler efficiency; heating system characteristics; the nature of controls and control schedules; R-values of building shell components; size and orientation of glass area; estimated air leakage rates; internal gains; local weather data (temperature and insolation); interior temperature measurements or estimates; hot-water consumption; plus lighting and other electrical loads. The program should be able to project the present monthly energy use by the building, which can then be compared with actual utility bills and weather data. A program should also be able to estimate the cost-effectiveness of retrofit options—for example, by ranking them according to estimated savings-to-investment (SIR) ratios (a discounted lifecycle cost index commonly used in weatherization work).

What follows is a brief review of a few leading computer auditing software packages suitable for use in apartment building retrofit programs. A broader list of software, some of which is applicable to apartment buildings, is given by Weiss and Brown (1989). Meier and Rainer (1991) discuss the principles of applied energy analysis software, and an updated discussion of some packages is given by Penn (1994).

CIRA

CIRA (Computerized Instrumented Residential Audit program) was developed in the early 1980s at Lawrence Berkeley Laboratory (LBL 1982; Sonderegger and Dixon 1983). CIRA was designed for modeling single family homes and was once described as "the most advanced residential energy analysis available today" (Commoner and Rodberg 1986). Among its attributes were (1) convenient flexible structure for data entry; (2) built-in descriptions for heating systems common to apartment buildings; (3) flexible structure for adding and modifying retrofits to the program's library of prescribed measures; (4) performance economic analysis and ranking of measures by cost-effectiveness. The program's disadvantages were an inability to model multiple zones and an assumption of thermostatic control. CIRA itself is no longer available as such, but several current audit programs—such as EA-QUIP, EEDO, and NEAT, described below— were developed as modified versions of CIRA and are based on its algorithms.

Robinson et al. (1986) provide an in-depth description of their use of the original CIRA (1.0) for planning and evaluating retrofits of 12 multifamily buildings in Minnesota. They modeled pre-retrofit consumption to within 5% of actual consumption, and the average CIRA-predicted savings of 28% closely matched the average measured savings of 31%. For individual buildings, the point spread between predicted and observed savings averaged 7.5%. Of particular note is the fact that these buildings differed greatly from one another in both pre-retrofit descriptions and the measures installed. This reference includes several memos from Gary Nelson explaining assumptions used in running the audit that may be valuable for guiding similar projects. Note that a good match between an audit program's predicted savings and the measured savings reflects the quality of the retrofit work as well as the potential accuracy of the program. In these Minnesota retrofits, performed by the Energy Resource Center (ERC) in St. Paul, there was a thorough audit and a careful implementation, with good quality control and follow-up.

EA-QUIP

EA-QUIP (Energy Auditing using the Queens Information Packet) was developed by Leonard Rodberg of Queens College, New York. It is a CIRA audit customized for typical low-income apartment buildings and adapted to run on IBM PC-type microcomputers. Subroutines were added to account for the physical condition of the boiler and distribution system and to compute energy losses due to system imbalances in portions of buildings that are overheated (Rodberg 1991). Other enhancements of EA-QUIP over CIRA 1.0 include

- Simplified and more flexible user-friendly procedures for data entry and management

- Algorithms for converting fuel delivery records to monthly consumption projections

- Apartment building adaptations and refinements for use of efficiency test data, part-load efficiency curves, tankless coil domestic hot-water loads, and treatment of solar gains

- Coverage of additional retrofit measures, such as overhauls and upgrades to boiler/burner systems, distribution system balancing, control and temperature monitoring systems, lighting retrofits, and air sealing between wall frames and windows

EA-QUIP is being actively updated and maintained, with many refinements being added beyond those noted here, including better modeling of retrofits as well as provisions to address nonenergy repairs plus safety and health measures.

EA-QUIP is complemented by ES-QUIP (Energy Savings Analysis using the Queens Information Package), a PC adaptation of PRISM for weather-normalized analysis of billing data. The result is a package with estimates of costs and benefits of various retrofit measures presented in the context of a building's historical consumption and findings in the field. In addition, EA-QUIP results based on audits of 20 New York tenements have been compiled in an "Action Guide" that can be used without access to the computer. EA-QUIP has been approved by the U.S. Department of Energy (DOE) for use in New York apartment weatherization programs. EA-QUIP has also been tested by SyrESCO on apartment buildings for both Weather Assistance Program (WAP) and private conservation efforts; it was found to provide a clear list of nonconflicting measures ranked by their paybacks (Thomas 1990). The main criticisms of EA-QUIP are that the retrofit measures are mostly limited to those appropriate for New York City housing stock and that there is no "back door" for users to easily expand the program to cover other situations. However, the program is written in BASIC, and source code is provided, so that auditors with programming skills can make their own modifications if desired.

EA-QUIP is public domain software, available upon request from Prof. Leonard Rodberg, Department of Urban Studies, Queens College, Flushing, NY 11367. Telephone: 718-997-5134.

EEDO

EEDO (Energy Economics Design Options) is an enhanced PC microcomputer version of CIRA using algorithms based on correlations with more complex mainframe programs, such as DOE-2.1. Its core, however, is still the original single-family version of CIRA, to which a more user-friendly interface has been added. EEDO accepts data inputs in one of three modes: Researcher, Utility Auditor, and House

Doctor, in decreasing levels of detail and complexity. For example, the Utility Auditor mode requires detailed information of sources of possible air leakage, whereas the House Doctor mode relies on a single blower-door measurement. Help menus and standard default values are available for most input elements. Calculations include active solar simulations based on the "F-chart" method and passive solar simulations based on the Los Alamos Solar Load Ratio method. Weather data are included for over 150 U.S. and Canadian cities. EEDO can generate six separate reports, ranging from basic energy use estimates to detailed payback results for proposed retrofits. Although the program has been around for quite some time now and is well marketed, few performance comparisons are available.

EEDO is available from Burt, Hill, Kosar, Rittlemann Associates (contact: Paul Scanlon, Vice President of Engineering), 4110 Morgan Center, Butler, PA 16001. Telephone: 412-285-4761. Pricing in 1994 was $395, with half price for multiple copies.

ASEAM

ASEAM (A Simplified Energy Analysis Method) is an energy auditing package developed for DOE by William Fleming & Associates. It is a personal computer program capable of handling complex buildings, modeling up to ten zones and 13 different types of heating, ventilation, and air conditioning (HVAC) systems. ASEAM can analyze monthly energy consumption, peak cooling and heating loads, and daylighting and is particularly valuable for its emphasis on HVAC and plant variables. It provides pull-down menus, default values from ASHRAE Standard 90.1P for use when data are unavailable, a parametric processor module allowing "what if" options analysis, lifecycle cost calculations, graphical presentations of results, and output convertible to spreadsheet formats. A particularly nice feature is the ability it gives the user to specify numerous combinations of retrofit measures that then can be analyzed by the computer overnight, yielding a comparative economic analysis of all permutations of options. Specific evaluations of ASEAM's performance in apartment building applications are unfortunately not available. A users' survey indicated that modeling results with ASEAM 2.1 were within 20% of metered results, depending on the auditor's ability to calibrate the model to the building.

As of this writing, the program's current version (ASEAM 3.0) is available free of charge to government agencies or federal contractors from Enterprise Advisory Services, Inc., Arlington, VA 22209. Telephone: 800-566-2877.

Other Software

The National Energy Audit, known as **NEAT**, is a CIRA-derived PC microcomputer package developed by Oak Ridge National Laboratory for use in DOE's Weatherization Assistance Program (ESTSC 1995). NEAT is designed mainly for single-family houses and smaller, town house–style multiunit buildings. The program features simplified data entry screens similar to the data collection forms used by weatherization auditors. It accepts measurement-based information for mechanical system efficiencies and infiltration rates but does not automate use of billing data to calibrate the model to the building. NEAT covers heating and cooling system conservation measures as well as shell retrofits, accounts for interactions among retrofit options, and provides cost-effectiveness rankings based on estimated savings-to-investment ratios (SIRs). NEAT is available from the Energy Science and Technology Software Center (ESTSC), Oak Ridge, TN 37831. Telephone: 615-576-2606; e-mail: estsc@adonis.osti.gov. The quoted price in early 1995 was $510 for general users and $250 for those working under a DOE contract.

Some audit software packages are promised to become publicly available but have not yet moved from being an in-house product for the developing organizations. One such package is the **Multifamily Energy Audit Program** developed by the Center for Neighborhood Technology (CNT) in Chicago and used for over six years by the Chicago Energy Savers Fund in auditing over 400 buildings (Buck 1989, 1990). Although the program was marketed nationally, few organizations obtained it. The major drawback to the Multifamily Energy Audit Program was the price: $3,960 plus $100 for each associate agency, including two days of training for eight people. Even now, CNT rarely uses the package in its own multifamily efforts, which have been greatly scaled back in recent years. Experience with the program and its underlying approach are described by Evens and Katrakis (1988) and Biederman and Katrakis (1989).

Another such package is the **WSES-ConCalc** (Weatherization Self Evaluation System, Consumption Calculator) system, two programs designed by the staff at SyrESCO in Syracuse, New York. This package was developed to assist weatherization auditors in making decisions about both single-family and apartment buildings. It was slated for public release in 1990 but never quite materialized. At present, the package is still being used in-house by SyrESCO, but its future as a commercial product is unclear after recent staff changes.

References

Abt Associates. 1989. *Housing Inspection Manual, Section 8 Existing Housing Program.* Washington, D.C.: U.S. Department of Housing and Urban Development, Office of Housing.

[AHS 1989] Bureau of the Census. 1991. *American Housing Survey for the United States in 1989.* H150/89. Washington, D.C.: U.S. Department of Commerce.

Altman, F. 1981. "Energy Conservation in Rental Housing: A Case Study of Minnesota's Residential Rental Retrofit Program." In *Multifamily Energy Conservation: A Reader,* edited by S. Kaye, 165–177. Hartford, Conn.: Coalition of Northeast Municipalities.

Apgar, W., Jr., G. Masnick, and N. McArdle. 1991. *Housing in America 1970–2000: The Nation's Housing Needs for the Balance of the 20th Century.* Cambridge: Harvard University, Joint Center for Housing Studies.

Ashmore, C. (U.S. Department of Housing and Urban Development). 1994. Personal communication to Sandra Nolden. February 1.

[ASHRAE] American Society of Heating, Refrigerating and Air-Conditioning Engineers. 1989. *Ventilation for Acceptable Indoor Air Quality.* Standard 62-1989. Atlanta, Ga.: American Society of Heating, Refrigerating and Air-Conditioning Engineers.

———. 1991. *HVAC Applications Handbook.* Atlanta, Ga.: American Society of Heating, Refrigerating and Air-Conditioning Engineers.

———. 1992. *HVAC Systems and Equipment Handbook.* Atlanta, Ga.: American Society of Heating, Refrigerating and Air-Conditioning Engineers.

———. 1994. *Energy Cost Allocation for Multiple Occupancy Residential Buildings.* Guideline 8-1994. Atlanta, Ga.: American Society of Heating, Refrigerating and Air-Conditioning Engineers.

[ASTM] American Society for Testing and Materials. 1983. *Test Method for Determining Air Leakage Rate by Tracer Dilution.* Standard E741. Philadelphia: American Society for Testing and Materials.

245

————. 1988. *Standard Test Method for Determining Air Leakage Rate by Fan Pressurization*. Standard E779. Philadelphia: American Society for Testing and Materials.

Austin, City of. 1994. *Multifamily Incentive Program: Summary of Rebate Items*. Austin, Tex.: Environmental and Conservation Services Department.

Baylon, D., and J. Heller. 1988. "Methodology and Results of Blower Door Testing in Small Multifamily Buildings." In *Proceedings of the ACEEE 1988 Summer Study on Energy Efficiency in Buildings*, 2:11–23. Washington, D.C.: American Council for an Energy-Efficient Economy.

Bellamy, J., and J. Fey. 1988. "Energy-Efficient New Buildings Through a Utility Service Standard." In *Proceedings of the ACEEE 1988 Summer Study on Energy Efficiency in Buildings*, 5:24–27. Washington, D.C.: American Council for an Energy-Efficient Economy.

Berkowitz, P. (Wisconsin Energy Conservation Corp., Madison). 1994. Personal communication to Sandra Nolden.

Berkowitz, P., and P. Newman. 1988. "Reinventing the WHEEL: An Integrated Approach to Energy Efficiency in the Rental Housing Sector." In *Proceedings of the ACEEE 1988 Summer Study on Energy Efficiency in Buildings*, 5:28–32. Washington, D.C.: American Council for an Energy-Efficient Economy.

Bernstein, S. (Center for Neighborhood Technology, Chicago). 1994. Personal communication to Sandra Nolden.

Bethke, J. (Urban Coalition Weatherization Program, Minneapolis). 1990. Personal communication.

Beyea, J., D. Harrje, and F. Sinden. 1977. "Energy Conservation in an Old 3-Story Apartment Complex." In *Proceedings of the Symposium on Energy Use Management*, edited by S. Fazzolare and W. Smith, 373–383. New York: Pergamon.

Biederman, N., and J.T. Katrakis. 1989. *Space Heating Improvements in Multifamily Buildings*. GRI/88-0111. Chicago: Gas Research Institute.

Bleviss, D. 1980. "Retrofitting Multifamily Housing." In *Proceedings of the ACEEE 1980 Summer Study on Energy Efficiency in Buildings*, 3:1–16. Washington, D.C.: American Council for an Energy-Efficient Economy.

Bleviss, D., and A. Gravitz. 1984. *Energy Conservation and Existing Rental Housing*. Washington, D.C.: Energy Conservation Coalition.

Brinch, J. 1995. *DOE-HUD Initiative on Energy Efficiency in Housing: A Federal Partnership*. Program Summary Report. ORNL/SUB/93-SM840V. Oak Ridge, Tenn.: Oak Ridge National Laboratory. August 15.

Brown, M., L. Berry, and L. Kinney. 1993. *Weatherization Works: An Interim Report of the National Weatherization Evaluation.* ORNL/CON-373. Oak Ridge, Tenn.: Oak Ridge National Laboratory.

Brown, M., and D. Beschen. 1992. "A Status Report on the National Weatherization Evaluation." In *Proceedings of the ACEEE 1992 Summer Study on Energy Efficiency in Buildings,* 7:27–28. Washington, D.C.: American Council for an Energy-Efficient Economy.

Bryan, D. 1995. "An Analysis of the Relative Impact of Individual Energy Conservation Measures on Multifamily Building Natural Gas Consumption." In *Proceedings of 1995 Energy Program Evaluation Conference.* Chicago: National Energy Program Evaluation Conference.

Buck, W. 1989. *Multifamily Energy Audit Program User's Manual, Version II.* Chicago: Center for Neighborhood Technology.

———. 1990. Promotional letter regarding the Multifamily Energy Audit Program. Chicago: Center for Neighborhood Technology.

Callaway, J., and A. Lee. 1988. "Evaluation of DOE's Partners in Low-Income Residential Retrofit Program: Summary Paper." In *Proceedings of the ACEEE 1988 Summer Study on Energy Efficiency in Buildings,* 5:37–41. Washington, D.C.: American Council for an Energy-Efficient Economy.

Cameron, L. (GRASP, Philadelphia). 1990. Personal communication to Tom Wilson.

Cason, T., R. Uhlaner, and P. Delaney. 1991. "Selling Residential Energy Efficiency: An Evaluation of Customer Acceptance Through Multiple Surveys and End-Use Metering." In *Energy Program Evaluation: Uses, Methods, and Results,* 218–224. Proceedings of the 1991 National Energy Program Evaluation Conference. Chicago.

[CCC] Citizens Conservation Corporation. 1994. Case studies of energy conservation performance contracting in multifamily housing. Unpublished manuscripts and technical reports. Boston: Citizens Conservation Corporation.

[CEE] Center for Energy and Environment. 1995. "Center for Energy and Environment Offers Low Interest Financing to Improve Your Property." Brochure. Minneapolis: Center for Energy and Environment.

Ciesielki, C.A. 1984. "The Role of Stagnation and Obstruction of Water Flow in Isolation of *Legionella pneumophila* from Hospital Plumbing." *Applied and Environmental Microbiology* 11:984–987, November.

ClimateMaster. 1994. "Park Chase Apartments—A HUD Project."

Writeup on ground source heat pump installation project. Oklahoma City: ClimateMaster, Inc. July 30.

[CNT] Center for Neighborhood Technology. 1992. *The Role of Community-Based Organizations in Demand Side Management.* Chicago: Center for Neighborhood Technology.

Colton, R.D., B. Sachs, and J. DeBarros. 1994. "Models of Public and Private Investment in Energy Efficiency for Low-Income Housing." In *Proceedings of the ACEEE 1994 Summer Study on Energy Efficiency in Buildings*, 6:39–46. Washington, D.C.: American Council for an Energy-Efficient Economy.

Commoner, B., and L. Rodberg. 1986. *Energy Conservation for New York City Low-Income Housing.* NYSERDA Report 87-9. Albany: New York State Energy Research and Development Authority.

Crockett, D. (Natural Resources Corp.). 1990. Personal communication.

Cummings, J., J. Tooley, and N. Moyer. 1993. *Duct Doctoring: Diagnosis and Repair of Duct System Leaks.* Cape Canaveral: Florida Solar Energy Center.

Davis, R. (Lighting Research Center, Troy, N.Y.). 1993. Personal communication to Loretta Smith.

DeCicco, J.M. 1988a. "Energy Conservation and Outdoor-Reset Control of Space Heating Systems." In *Proceedings of the ACEEE 1988 Summer Study on Energy Efficiency in Buildings*, 2:24–37. Washington, D.C.: American Council for an Energy-Efficient Economy.

———. 1988b. *Modeling, Diagnosis, and Implications for Improving the Energy Efficiency of Centrally-Heated Apartment Buildings.* Ph.D. dissertation and Report PU/CEES No. 225. Princeton, N.J.: Princeton University, Center for Energy and Environmental Studies.

DeCicco, J.M., and G. Dutt. 1986. "Domestic Hot Water Service in Lumley Homes: A Comparison of Energy Audit Diagnosis with Instrumented Analysis." In *Proceedings of the ACEEE 1986 Summer Study on Energy Efficiency in Buildings*, 2:33–46. Washington, D.C.: American Council for an Energy-Efficient Economy.

DeCicco, J.M., and W. Kempton. 1986. "Heating a Multifamily Building: Tenant Perceptions and Behavior." In *Proceedings of the ACEEE 1986 Summer Study on Energy Efficiency in Buildings*, 7:47–61. Washington, D.C.: American Council for an Energy-Efficient Economy.

———. 1987. "Behavioral Determinants of Energy Consumption in a Centrally Heated Apartment Building." *Energy Systems and Policy* 2:155–168.

Diamond, R.C. 1986. "A Case Study of the Determinants of Energy Use in Housing for the Low-Income Elderly." Ph.D. dissertation.

Berkeley: University of California, Department of Architecture. April.

Diamond, R.C., C. Goldman, M.P. Modera, M. Rothkopf, M. Sherman, and E.L. Vine. 1985. *Building Energy Retrofit Research, Multifamily Sector*. LBL-20165. Berkeley: Lawrence Berkeley Laboratory.

Diamond, R.C., J.A. McAllister, L.I. Rainer, R.L. Ritschard, H.E. Feustel. 1992. "Affordable Housing Rehabilitation in Vermont: Energy Savings, Cost-Effectiveness, and Resident Satisfaction." In *Proceedings of Thermal Performance of the Exterior Envelopes of Buildings V*, 345–354. Clearwater, Fla. December.

Diamond, R.C., M.P. Modera, and H.E. Feustel. 1986. "Ventilation and Occupant Behavior in Two Apartment Buildings." In *Proceedings of the 7th Air Infiltration Centre Conference on Occupant Interaction with Ventilation Systems*, 6.1–6.18. Stratford, England. September.

Dixon, W. 1992. "Energy-Efficient Windows." *Construction Specifier*. October: 76–77.

[DOE] U.S. Department of Energy. 1979. *Achieving Energy Conservation in Existing Apartment Buildings*. Washington, D.C.: Office of Buildings and Community Services.

————. 1984. *Multifamily Buildings: A Draft Technology Transfer Analysis and Plan*. Washington, D.C.: Office of Buildings and Community Systems.

————. 1985. *Building Energy Retrofit Research: Multifamily Sector, Multiyear Plan FY 1986–1991*. DOE/CE-0142. Washington, D.C.

————. 1994. *FY 1995 Program Plan for Climate Change Action No. 1: Rebuild America*. Washington, D.C.: Office of Energy Efficiency and Renewable Energy. December 14.

Domus Plus. 1994. "Energy-Efficient Affordable Housing Program." Report to Illinois Department of Energy and Natural Resources. Oak Park, Ill.: Domus Plus. September.

Dutt, G., and D. Harrje. 1989. *Multifamily Building Energy Diagnostics Technical Reference Manual*. Princeton, N.J.: Princeton University, Center for Energy and Environmental Studies.

Dyballa, C., and C. Connelly. 1992. "Electric and Water Utilities: Building Cooperation and Savings." In *Proceedings of the ACEEE 1992 Summer Study on Energy Efficiency in Buildings*, 5:51–61. Washington, D.C.: American Council for an Energy-Efficient Economy.

Eckman, T., N. Benner, and F.M. Gordon. 1992. "It's 2002: Do You Know Where Your Demand-Side Management Policies and Programs Are?" In *Proceedings of the ACEEE 1992 Summer Study on Energy Efficiency in Buildings*, 5:1–17. Washington, D.C.: American Council for an Energy-Efficient Economy.

Englander, S., and G. Dutt. 1986. "DHW Energy Savings in Multifamily Buildings." In *Proceedings of the ACEEE 1986 Summer Study on Energy Efficiency in Buildings*, 1:47–61. Washington, D.C.: American Council for an Energy-Efficient Economy.

[EPRI] Electric Power Research Institute. 1995. "Study of *Legionella* in Electric Water Heaters Reported." *Environment Update* 9(2): 5. Palo Alto, Calif.: Electric Power Research Institute. May.

[ESTSC] Energy Science and Technology Software Center. 1995. *National Energy Audit (NEAT), Software Abstract*. Oak Ridge, Tenn.: Energy Science and Technology Software Center.

Evens, A., and J.T. Katrakis. 1988. "Chicago's Residential Energy Conservation Loan Fund: A Preliminary Evaluation of Its Impact on Multifamily Buildings." In *Proceedings of the ACEEE 1988 Summer Study on Energy Efficiency in Buildings*, 2:38–50. Washington, D.C.: American Council for an Energy-Efficient Economy.

Ewing, G., D. Neumeyer, S. Pigg, and J. Schlegel. 1988. "Effectiveness of Boiler Control Retrofits on Multifamily Buildings in Wisconsin." In *Proceedings of the ACEEE 1988 Summer Study on Energy Efficiency in Buildings*, 2:51–56. Washington, D.C.: American Council for an Energy-Efficient Economy.

Fay, B. 1984. "Voluntary Rental Living Unit Program." In *Proceedings of the ACEEE 1984 Summer Study on Energy Efficiency in Buildings*, C:18–29. Washington, D.C.: American Council for an Energy-Efficient Economy.

Feist, J., R. Farhang, J. Erikson, E. Stergakos, P. Brodie, and P. Liepe. 1994. "Super Efficient Refrigerators: The Golden Carrot from Concept to Reality." In *Proceedings of the ACEEE 1994 Summer Study on Energy Efficiency in Buildings*, 3:67–75. Washington, D.C.: American Council for an Energy-Efficient Economy.

Fels, M.F. 1986. "PRISM: An Introduction." In *Measuring Energy Savings: The Scorekeeping Approach*, edited by M.F. Fels. Special Issue of *Energy and Buildings* 9(1&2): 5–18.

Fels, M.F., J. Kissock, and M. Marean. 1994. "Model Selection Guidelines for PRISM." In *Proceedings of the ACEEE 1994 Summer Study on Energy Efficiency in Buildings*, 8:49–61. Washington, D.C.: American Council for an Energy-Efficient Economy.

Fels, M.F., and C.L. Reynolds. 1993. *Energy Analysis in New York City Multifamily Buildings*. NYSERDA Report 93-3. Albany: New York State Energy Research and Development Authority.

Fenichel, A. 1992. "Low-Income Weatherization: DSM Resource, Customer Service, or Social Equity Program?" *DSM Quarterly*. Fall: 16–25.

Ferrey, S. 1986. "Financing Energy Efficiency in Public Housing: Innovative Regulatory Opportunities." In *Proceedings of the ACEEE*

1986 Summer Study on Energy Efficiency in Buildings, 4:52–64. Washington, D.C.: American Council for an Energy-Efficient Economy.

Feustel, H.E., C. Zuercher, R.C. Diamond, J.B. Dickinson, D.T. Grimsrud, and R.D. Lipschutz. 1985. "Air Flow Pattern in a Shaft-Type Structure." *Energy and Buildings* 8(2): 105–122.

Fitzgerald, J. 1989. "Blown Cellulose as Air Leakage Control." *Energy Exchange* 1(2).

———. (Fitzgerald Contracting, Minneapolis). 1990. Personal communication to Tom Wilson.

———. 1993. Personal communication to Tom Wilson.

Fitzgerald, J., G. Nelson, and L. Shen. 1990. "Sidewall Insulation and Air Leakage Control." *Home Energy* 7(1): 13–20.

Francisco, P.W., and L. Palmiter. 1994. "Infiltration and Ventilation Measurements on Three Electrically Heated Multifamily Buildings." In *Proceedings of the ACEEE 1994 Summer Study on Energy Efficiency in Buildings*, 5:97–104. Washington, D.C.: American Council for an Energy-Efficient Economy.

Freedberg, M., and D. Schumm. 1986. "New Initiatives in Financing Multifamily Energy Conservation: Recent Developments in Chicago." In *Proceedings of the ACEEE 1986 Summer Study on Energy Efficiency in Buildings*, 4:65–75. Washington, D.C.: American Council for an Energy-Efficient Economy.

Fronczeck, P., and H. Savage. 1995. *1991 Residential Finance Survey*. Report CH-4-1. Washington, D.C.: U.S. Department of Commerce, Bureau of the Census. March.

Garland, M. 1986. "Performance Contracting: Planning for Change in Taxes, Regulations, and the Market." In *Financing Energy Conservation*, edited by M. Weedall, R. Weisenmiller, and M. Shepard, 99–110. Washington, D.C.: American Council for an Energy-Efficient Economy.

Gill-Polley, Lydia. (Constructive Consulting, Inc., Wichita, Kans.). 1990. Personal communication to Tom Wilson.

Gold, C. 1984. "The Page Homes Demonstration Energy Conservation Computer System." In *What Works: Documenting Energy Conservation in Buildings*, edited by J. Harris and C. Blumstein, 140–150. Washington, D.C.: American Council for an Energy-Efficient Economy.

Goldman, C., K. Greely, and J. Harris. 1988a. "Retrofit Experience in the U.S. Multifamily Buildings: Energy Savings, Costs, and Economics." *Energy* 13(11): 797–811.

———. 1988b. *Retrofit Experience in U.S. Multifamily Buildings: Energy Savings, Costs, and Economics, Volume I*. Report LBL-25248

(1/2). Berkeley: Lawrence Berkeley Laboratory, Applied Science Division.

————. 1988c. *Retrofit Experience in the U.S. Multifamily Buildings: Energy Savings, Costs, and Economics, Volume II.* Report LBL-25248(1/2). Berkeley: Lawrence Berkeley Laboratory, Applied Science Division.

————. 1988d. "An Updated Compilation of Measured Energy Savings in Retrofitted Multifamily Buildings." In *Proceedings of the ACEEE 1988 Summer Study on Energy Efficiency in Buildings,* 2:75–88. Washington, D.C.: American Council for an Energy-Efficient Economy.

Goldman, C., and R. Ritschard. 1986. "Energy Conservation in Public Housing: A Case Study of the San Francisco Housing Authority." *Energy and Buildings* 9:89–98.

Goldner, F. 1992. "Multifamily Building Energy Monitoring and Analysis, Domestic Hot Water Use and System Sizing Criteria Development: A Status Report." In *Proceedings of the ACEEE 1988 Summer Study on Energy Efficiency in Buildings,* 2:75–88. Washington, D.C.: American Council for an Energy-Efficient Economy.

————. 1994. *Energy Use and Domestic Hot Water Consumption.* NYSERDA Report 94-19. Albany: New York State Energy Research and Development Authority. November.

Goodman, J.L., and M.R. Gurpe. 1995. "Top Ten Surprises About Ownership and Financing of Rental Housing." *Real Estate Finance.* Winter.

Graham, A., S. Pigg, J. Schlegal, and P. Harrison. 1991. "An Impact Evaluation of a Multifamily Gas Conservation Loan Program in Chicago: Methods and Results." In *Energy Program Evaluation: Uses, Methods, and Results,* 408–415. Proceedings of the 1991 National Energy Program Evaluation Conference. Chicago.

Greely, K., and C. Goldman. 1989. "Retrofit Experience in U.S. Multifamily Buildings." *Home Energy* 6(2): 16–20.

Greely, K., C. Goldman, and R. Ritschard. 1986. "Analyzing Energy Conservation Retrofits in Public Housing: Savings, Cost-Effectiveness, and Policy Implications." In *Proceedings of the ACEEE 1986 Summer Study on Energy Efficiency in Buildings,* 2:125–137. Washington, D.C.: American Council for an Energy-Efficient Economy.

Greely, K., C. Goldman, R. Ritschard, and M. Jackson. 1987. *Baseline Analysis of Measured Energy Consumption in Public Housing.* LBL-22854. Berkeley: Lawrence Berkeley Laboratory.

Green, R., and B. Grande. 1992. "Filling the Gap." *The Construction Specifier.* October: 48.

Griffin, T., J. Kerwin, and S. Thompson. 1984. "An Energy Management Service for Multifamily Rental Property in St. Paul, Minnesota." In *Proceedings of the ACEEE 1984 Summer Study on Energy Efficiency in Buildings*, C:30–39. Washington, D.C.: American Council for an Energy-Efficient Economy.

Groberg, R. (U.S. Department of Housing and Urban Development). 1994. Personal communication to Sandra Nolden.

———. 1995. Personal communication to John DeCicco.

Guyton, M. 1993. "Measured Performance of Relocating Air Distribution Systems in an Existing Residential Building." *ASHRAE Transactions* 99(2).

Haber, E.A. 1994. "Affordable Housing—Moving Beyond Energy Conservation." In *Affordable Comfort 1994: Selected Readings*, 27–30. Evanston, Ill.: Affordable Comfort Conference. March.

Hackett, B. 1984. "Energy Billing Systems and the Social Control of Energy Use in a California Apartment Complex." In *Families and Energy: Coping with Uncertainty*, edited by B.M. Morrison and W. Kempton, 291–301. Lansing: Michigan State University, Institute for Family and Child Study.

Hamilton, L., K. Tohinaka, and C. Neme. 1994. "New Smart Protocols to Avoid Lost Opportunities and Maximize Impact of Residential Retrofit Programs." In *Proceedings of the ACEEE 1994 Summer Study on Energy Efficiency in Buildings*, 9:147–157. Washington, D.C.: American Council for an Energy-Efficient Economy.

Hammarlund, J., J. Proctor, G. Kast, and T. Ward. 1992. "Enhancing the Performance of HVAC and Distribution Systems in Residential New Construction." In *Proceedings of the ACEEE 1992 Summer Study on Energy Efficiency in Buildings*, 2:85–87. Washington, D.C: American Council for an Energy-Efficient Economy.

Harris, J., A. Rosenfeld, C. Blumstein, and J. Millhone. 1992. "Creating Institutions for Energy Efficiency R&D: New Roles for States and Utilities." In *Proceedings of the ACEEE 1992 Summer Study on Energy Efficiency in Buildings*, 6:91–102. Washington, D.C.: American Council for an Energy-Efficient Economy.

Harrje, D. 1984. "An Introduction to the Use of Advanced Analysis Tools for Effective Weatherization." In *The Instrumented Audit*, edited by T. Wilson and R. Belshe. Transcriptions of presentations at the 1984 Weatherization Conference, French Lick Springs, Indiana. Indianapolis: Indiana Community Action Program Directors Association.

Hasterok, L. 1990. "Multifamily Lighting: Come On In, the Savings Are Fine!" *Home Energy* 7(6): 12–16.

Hayes, V. 1992. "Auditing the All-Electric Multifamily Building." *Home Energy* 9(5): 15–19.

Hayes, V. 1995. (National Center for Appropriate Technology, Butte, Mont.). Personal communication to John DeCicco. December.

Hemphill, M. 1981. "Marketing and Delivery of Multifamily Conservation Programs in Portland, Oregon." In *Multifamily Energy Conservation: A Reader*, edited by S. Kaye, 153–162. Hartford, Conn.: Coalition of Northeast Municipalities.

Henderson, B.H. 1992. Park Chase Apartments Thermal Improvement Proposal. Good Cents Energy-Efficient Apartment Program, Public Service Company of Oklahoma, Tulsa. March 4.

Hewett, M.J. 1988a. "Billing Tenants for Heat: Paying for What You Use." *Home Energy* 5(1): 10–17.

———. 1988b. "Multifamily Building Technologies: Panel Overview." In *Proceedings of the ACEEE 1988 Summer Study on Energy Efficiency in Buildings*, 2:7–10. Washington, D.C.: American Council for an Energy-Efficient Economy.

———. 1988c. "Outdoor Resets and Cutouts: Quick Fixes for Hot-Water Heaters." *Home Energy* 5(5): 15–20.

———. 1990. "Do Vent Dampers Work in Multifamily Buildings?" *Home Energy* 7(2): 27–32.

———. 1991. "Resets and Cutouts: Still a Good Idea." *Home Energy* 8(5): 42–43.

Hewett, M.J., H. Emslander, and M. Koehler. 1986. "Heating Cost Allocation in Centrally Heated Rental Housing: Energy Conservation Potential and Standards Issues." In *Proceedings of the ACEEE 1986 Summer Study on Energy Efficiency in Buildings*, 2:141–161. Washington, D.C.: American Council for an Energy-Efficient Economy.

Hewett, M.J., M.S. Lobenstein, and S. Nathan. 1994. "The Science and Art of Retrofitting Low-Rise Multifamily Buildings." *ASHRAE Journal* 36(5): 41–49.

Hewett, M.J., and G. Peterson. 1984. "Measured Energy Savings from Outdoor Resets in Modern, Hydronically Heated Apartment Buildings." In *Proceedings of the ACEEE 1984 Summer Study on Energy Efficiency in Buildings*, 3:135–152. Washington, D.C.: American Council for an Energy-Efficient Economy.

Hewitt, D., D. Palermini, J. Pratt, and D. Tooze. 1988. "Understanding What Motivates Multifamily Property Owners." In *Proceedings of the ACEEE 1988 Summer Study on Energy Efficiency in Buildings*, 7:57–69. Washington, D.C.: American Council for an Energy-Efficient Economy.

[HHS] U.S. Department of Health and Human Services. 1995. *Adminis-

tration for Children and Families, Justification of Estimates for Appropriations Committees, Fiscal Year 1995. Washington, D.C.

Hickman, C., and T. Steele. 1991. "Building Site Visits as Supplement to Program Evaluation." In *Energy Program Evaluation: Uses, Methods, and Results,* 174–180. Proceedings of the 1991 National Energy Program Evaluation Conference. Chicago.

Hirst, E., R. Goeltz, and M. Hubbard. 1987. "Determinants of Electricity Use for Residential Water Heating: The Hood River Conservation Project." *Energy Conversion and Management* 27(2).

Home Energy. 1993. Special Issue, "Ducts Rediscovered: A Guide for Auditors, Contractors, and Utilities." *Home Energy* 10(5).

———. 1995. Special Issue on Multifamily Buildings. *Home Energy* 12(5).

Hubinger, G. 1984. "Improving the Regulatory Approach to Increased Rental Housing Energy Efficiency: The Minnesota Case Study." In *Proceedings of the ACEEE 1984 Summer Study on Energy Efficiency in Buildings,* 3:68–82. Washington, D.C.: American Council for an Energy-Efficient Economy.

[HUD] U.S. Department of Housing and Urban Development. 1986. *Solar Energy and Energy Conservation Bank Annual Report to Congress for FY 1986.* Washington, D.C.

Humburgs, C.L. 1993. *Evaluation of Multifamily Common-Area Lighting in the Energy Smart Design Program.* Seattle: Seattle City Light, Energy Management Services Division. February.

Hunt, A. (Vermont Housing Finance Agency). 1993. Personal communication to Janice DeBarros.

Illinois [DCCA] Department of Commerce and Community Affairs, Weatherization Assistance Program. 1992. *Multifamily Building Plan.* Springfield.

———. 1994. *Multifamily Building Assessment; Apartment Unit Worksheet; Apartment Building Heating System Assessment.* Springfield.

Jacobson, B.B., S.S. Ellison, M.L. Gallicchio, A.L. Bachman, and F.M. Gordon. 1988. "Demand Management Development Decision Matrix for Low-Income/Special Needs Customers: A Program Ranking and Marketing Tool." In *Proceedings of the ACEEE 1988 Summer Study on Energy Efficiency in Buildings,* 6:89–100. Washington, D.C.: American Council for an Energy-Efficient Economy.

Jacobson, D., M. Miller, C. Granda, D. Conant, R. Wright, and D. Landsberg. 1992. "Comparing Engineering Estimates to Measured Savings: One Utility's Experience." In *Proceedings of the ACEEE 1992 Summer Study on Energy Efficiency in Buildings,* 7:109–119. Washington, D.C.: American Council for an Energy-Efficient Economy.

[JCHS] Joint Center for Housing Studies. 1992. *The State of the Nation's Housing*. Cambridge: Harvard University.

———. 1993. *The State of the Nation's Housing*. Cambridge: Harvard University.

Jordan, B. 1990. "Door-to-Door Water Conservation Retrofits: The San Jose Story." *Home Energy* 7(4): 25–31.

Judd, P. 1993. "The Overheated City: Accountability, Not Technology, Conserves Energy." *Strategic Planning for Energy and the Environment* 13(2): 34–43.

Kamalay, L. 1992. "Conservation and the Affordable Home." *Construction Specifier*. October: 74–85.

———. (Citizens Conservation Corporation, Boston). 1994. Personal communication to John DeCicco.

Karins, N. (New York State Energy Research and Development Authority, Albany). 1994. Personal communication to John DeCicco.

Katrakis, J.T. 1989a. *Demonstration of Comprehensive Retrofits of Multifamily Buildings in the Low-Income Weatherization Program: Final Report*. Chicago: Center for Neighborhood Technology.

———. 1989b. "Retrofitting Single-Pipe Steam Heating Systems." *Home Energy* 6(6): 16–21.

———. 1990. "Effects of Tenant Comfort on Retrofit Performance in Older Low-Income Multifamily Buildings." In *Proceedings of the ACEEE 1990 Summer Study on Energy Efficiency in Buildings*, 9:133–142. Washington, D.C.: American Council for an Energy-Efficient Economy.

Katrakis, J.T., B. Burris, T. Catlin, and D. Fish. 1986. "Reducing Temperature Imbalance in Single-Pipe Steam Buildings." In *Proceedings of the ACEEE 1986 Summer Study on Energy Efficiency in Buildings*, 2:173–188. Washington, D.C.: American Council for an Energy-Efficient Economy.

Katrakis, J.T., P.A. Knight, and J.D. Cavallo. 1994. *Energy-Efficient Rehabilitation of Multifamily Buildings in the Midwest*. Report ANL/DIS/TM-16, Argonne National Laboratory. Argonne, Ill. September.

Katrakis, J.T., L. Wharton, and W. Goldman. 1992. "Choosing the Right Replacement Boiler in Low-to-Moderate Income Multifamily Buildings: An Update of Current Practice and Research." In *Proceedings of the ACEEE 1992 Summer Study on Energy Efficiency in Buildings*, 2:117–126. Washington, D.C.: American Council for an Energy-Efficient Economy.

Katrakis, J.T., and T.S. Zawacki. 1993. "How to Measure Low-Pressure Steam Boiler Efficiency." *ASHRAE Journal* 35(9): 46–55.

Kelly, M., J. McQuail, and R. O'Brien. 1992. "Case Study of Infiltration and Ventilation Improvements in a Multifamily Building." In *Proceedings of the ACEEE 1992 Summer Study on Energy Efficiency in Buildings*, 2:127–133. Washington, D.C.: American Council for an Energy-Efficient Economy.

Kempton, W. 1986. "Two Theories of Home Heat Control." *Cognitive Science* 10:75–90.

Kempton, W., and J.M. DeCicco. 1985. *Residents' Perceptions of Heating and Ventilation in a Senior Citizen Public Housing Project*. Working Paper No. 80. Princeton, N.J.: Princeton University, Center for Energy and Environmental Studies. December.

Kempton, W., D. Feuermann, and A. McGarity. 1989. *"I Always Turn It On Super": User Conceptions and Operation of Room Air Conditioners*. Princeton, N.J.: Princeton University, Center for Energy and Environmental Studies.

Keppler, M. 1978. "Energy Tax Act of 1978 Brings New Tax Credits for Individuals and Businesses." *Journal of Taxation*. December.

Kinney, L. (Synertech Systems Corp., Syracuse). 1994. Personal communication to Sandra Nolden.

Knight, P.A. (Domus Plus, Oak Park, Ill.). 1993–1995. Personal communications to John DeCicco.

Kushler, M., K. Keating, J. Schlegel, and E.L. Vine. 1992. "The Purpose, Practice, and Profession of DSM Evaluation: Current Trends, Future Challenges." In *Proceedings of the ACEEE 1992 Summer Study on Energy Efficiency in Buildings*, 7:1–25. Washington, D.C.: American Council for an Energy-Efficient Economy.

Lambert, F. 1986. "Innovative Financing for Energy Conservation in Buildings." In *Proceedings of the ACEEE 1986 Summer Study on Energy Efficiency in Buildings*, 10:186–189. Washington, D.C.: American Council for an Energy-Efficient Economy.

Landry, R.W., D.E. Maddox, M.S. Lobenstein, and D.L. Bohac. 1993. "Measuring Seasonal Efficiency of Space Heating Boilers." *ASHRAE Journal* 35(9): 38–45.

[LBL] Lawrence Berkeley Laboratory. 1982. *CIRA Reference Manual*. EEB-EPB Pub-442. Berkeley: Lawrence Berkeley Laboratory, Energy Efficient Buildings Program.

L'Ecuyer, M., H.M. Sachs, G. Fernstrom, D. Goldstein, E.C. Klumpp, and S. Nadel. 1992. "Stalking the Golden Carrot: A Utility Consortium to Accelerate the Introduction of Super-Efficient, CFC-Free Refrigerators." In *Proceedings of ACEEE 1992 Summer Study on Energy Efficiency in Buildings*, 5:137–145. Washington, D.C.: American Council for an Energy-Efficient Economy.

[LIHIS] Low Income Housing Information Service. 1995. *Out of Reach: Why Everyday People Can't Find Affordable Housing.* Low Income Housing Information Service. Washington, D.C. March.

Lines, J., and M. O'Brien. 1994. "Energy Performance Contracting for Subsidized Housing." *NAHMA News* 5(4). Alexandria, Va.: National Assisted Housing Managers Association.

Lipschutz, R.D., R.C. Diamond, and R. Sonderegger. 1983. *Energy Use in a High-Rise Apartment Building.* LBL-16366. Berkeley: Lawrence Berkeley Laboratory.

Lobenstein, M.S. 1989. "Boiler Tune-up: Improving the 'MPG' of Multifamily Buildings." *Home Energy* 6(5): 21–26.

———. (Center for Energy and Environment, Minneapolis). 1994-95. Personal communications to John DeCicco.

Lobenstein, M.S., D.L. Bohac, T.S. Dunsworth, and R.W. Landry. 1990. "Measured Savings and Field Experience from the Installation of Front-End Modular Boilers." In *Proceedings of the ACEEE 1990 Summer Study on Energy Efficiency in Buildings,* 9:189–201. Washington, D.C.: American Council for an Energy–Efficient Economy.

Lobenstein, M.S., D.L. Bohac, K. Korbel, M. Hancock, T.S. Dunsworth, and T. Staller. 1992. "Field Testing of Various Energy Saving Measures for Domestic Hot Water in Multifamily Buildings." In *Proceedings of the ACEEE 1992 Summer Study on Energy Efficiency in Buildings,* 2:145–155. Washington, D.C.: American Council for an Energy-Efficient Economy.

Lobenstein, M.S., D.L. Bohac, T. Staller, M. Hancock, and T.S. Dunsworth. 1992. *Measured Savings from Integral Flue Dampers and Thermal Vent Dampers on Domestic Hot Water Heaters in Multifamily Buildings.* Minneapolis: Center for Energy and Environment.

Lobenstein, M.S., and T.S. Dunsworth. 1989. *Converting Two Pipe Steam Heated Buildings to Hot Water Heat: Measured Savings and Field Experience.* Minneapolis: Minneapolis Energy Office.

———. 1990. "Measured Savings from Converting Two Pipe Steam Buildings to Hot Water Heat." In *Proceedings of the ACEEE 1990 Summer Study on Energy Efficiency in Buildings,* 9:185–187. Washington, D.C.: American Council for an Energy-Efficient Economy.

Lobenstein, M.S., M.J. Hewett, and T.S. Dunsworth. 1986. "Converting Steam Heated Buildings to Hot Water Heat: Practices, Savings and Other Benefits." In *Proceedings of the ACEEE 1986 Summer Study on Energy Efficiency in Buildings,* 1:140–154. Washington, D.C.: American Council for an Energy-Efficient Economy.

———. 1993. "Energy Savings and Field Experience from Converting

Steam-Heated Buildings to Hydronic Heat." CH-93-15-3. *ASHRAE Transactions* 99(1).

Lobenstein, M.S., and G. Peterson. 1988. "Evaluating Options for the Conversion of Two Pipe Steam Heating Systems to Hot Water Heat: A Bidding and Specification Guide." In *Proceedings of the ACEEE 1988 Summer Study on Energy Efficiency in Buildings,* 2:132–147. Washington, D.C.: American Council for an Energy-Efficient Economy.

Lubke, T. 1994. "Utility Allowances in Public and Assisted Housing." Policy Analysis Paper. Cambridge: Harvard University, John F. Kennedy School of Government.

McBride, J., S. Thomas, and D. Valk. 1990. "End-Use Energy Consumption in Small Multifamily Buildings." In *Proceedings of the ACEEE 1990 Summer Study on Energy Efficiency in Buildings,* 9:207–216. Washington, D.C.: American Council for an Energy-Efficient Economy.

McClelland, L. 1983. *Tenant Paid Energy Costs in Multifamily Rental Housing: Effects on Energy Use, Owner Investment and Market Value of Energy.* Boulder: University of Colorado, Institute of Behavioral Science.

MacDonald, J.M. 1993. *Description of the Weatherization Assistance Program in Larger Multifamily Buildings for Program Year 1989.* Report ORNL/CON-329. Oak Ridge, Tenn.: Oak Ridge National Laboratory. April.

Manheimer, B. 1992. "Performance Contracting in Public Housing." *Construction Specifier.* October: 82–83.

Marsh Technical Services. 1991. *Estimates of Energy Savings from Weatherization of Multifamily Dwellings.* Portland, Ore.: Portland Energy Office.

Masker, J. (National Center for Appropriate Technology, Washington, D.C.). 1995. Personal communication to John DeCicco. May.

Mast, B., and B. Ignelzi. 1994. "The Role of Incentives and Information in DSM Programs." In *Proceedings of the ACEEE 1994 Summer Study on Energy Efficiency in Buildings,* 10:145–153. Washington, D.C.: American Council for an Energy-Efficient Economy.

[MDPS] Minnesota Department of Public Service. 1987. *Multifamily Building Energy Audit, Technical Manual and Multifamily Building Energy Auditor Training Program Resource Materials.* Prepared by the Energy Division and the Center for Local Self-Reliance, Minneapolis.

Meier, A., and L.I. Rainer. 1991. "Computing Energy Savings: A Software Overview." *Home Energy* 8(5): 13–18.

Mills, E., M.F. Fels, and C.L. Reynolds. 1987. "PRISM: A Tool for Tracking Retrofit Savings." *Energy Auditor & Retrofitter* 4(6): 27–36.

Mills, E., R. Ritschard, and C. Goldman. 1986. "Financial Impacts of Energy Conservation Investment in Public Housing." In *Proceedings of the ACEEE 1986 Summer Study on Energy Efficiency in Buildings*, 4:119–131. Washington, D.C.: American Council for an Energy-Efficient Economy.

Modera, M.P. (Lawrence Berkeley Laboratory). 1993. Personal communication to John DeCicco.

Modera, M.P., J.T. Brunsell, and R.C. Diamond. 1985. "Improving Diagnostics and Energy Analysis for Multifamily Buildings: A Case Study." In *Proceedings of Thermal Performance of the Exterior Envelopes of Buildings*, 689–706. Clearwater, Fla. December.

Morgan, S. 1994. "Utility Multifamily DSM Programs and Public and Assisted Housing: Fitting Program Designs to Energy Opportunities." In *Proceedings of the ACEEE 1994 Summer Study on Energy Efficiency in Buildings*, 10:181–186. Washington, D.C.: American Council for an Energy-Efficient Economy.

Morrill, J. (American Council for an Energy-Efficient Economy, Washington, D.C.). 1995. Personal communication to John DeCicco.

Nadel, S. 1990. "Electrical Utility Conservation Programs: A Review of the Lessons Taught by a Decade of Program Experience." In *Proceedings of the ACEEE 1990 Summer Study on Energy Efficiency in Buildings*, 8:181–205. Washington, D.C.: American Council for an Energy-Efficient Economy.

National Low Income Housing Preservation Commission. 1988. *Preventing the Disappearance of Low-Income Housing.* Report to the House Subcommittee on Housing and Community Development. Washington, D.C.

Newcomb, J. 1994. *The Future of Energy Efficiency Services in a Competitive Environment.* Strategic Issues Paper No. FIP-IV. Boulder: E-Source, Inc. May.

New York State Energy Office. Undated. *Multifamily Housing Energy Conservation Workbook.* Albany.

[NFRC] National Fenestration Rating Council. 1993. *Certified Product Directory.* Silver Spring, Md.

Nolden, S., and J. Snell. 1995. *Persistence of Savings from Equipment Retrofits in Low-Income Apartment Buildings.* Report prepared for Oak Ridge National Laboratory under the DOE-HUD Initiative. Boston: Citizens Conservation Corporation. April.

Norton, R., and M. Lindberg. 1992. "Highlights of Recent Projects of the Urban Consortium Energy Task Force." In *Proceedings of the ACEEE 1992 Summer Study on Energy Efficiency in Buildings*, 6:189–192. Washington, D.C.: American Council for an Energy-Efficient Economy.

[NYCHA] New York City Housing Authority. 1982. *Housing Fireman's Guide*. New York: New York City Housing Authority. January.

[NYC/HPD] New York City Department of Housing Preservation and Development, Energy Conservation Division. Undated. *Aluminum Thermal Windows: Guides to Installation, Maintenance, and Repair*. New York.

[NYSERDA] New York State Energy Research and Development Authority. 1995. *Simplified Multizone Blower Door Techniques for Multifamily Buildings*. NYSERDA Report 95-16. Prepared by Steve Winter Associates, Inc., with Lawrence Berkeley Laboratory, Energy Performance Group. Albany: New York State Energy Research and Development Authority. September.

O'Keefe, L. (Portland Energy Office). 1994. Personal communication to Sandra Nolden. June 23.

Okumo, D. 1990. "Multifamily Retrofit Electricity Savings: The Seattle City Light Experience." In *Proceedings of the ACEEE 1990 Summer Study on Energy Efficiency in Buildings*, 6:119–130. Washington, D.C.: American Council for an Energy-Efficient Economy.

———. 1991. *The Multifamily Conservation Program: Evaluation of Electricity Savings and Costs*. Seattle: Seattle City Light.

———. 1992. "Revisiting Multifamily Retrofit Electricity Savings." In *Proceedings of the ACEEE 1992 Summer Study on Energy Efficiency in Buildings*, 7:165–173. Washington, D.C.: American Council for an Energy-Efficient Economy.

Okumo Tachibana, D., ed. 1993. *Energy Conservation Accomplishments: 1977–1992*. Seattle: Seattle City Light, Energy Management Services Division.

——— (Seattle City Light). 1994. Personal communication to Sandra Nolden. January 25.

[ORNL] Oak Ridge National Laboratory. 1992. *Energy Performance Contracting for Public and Indian Housing: A Guide for Participants*. Report prepared for the Energy Division of the U.S. Department of Housing and Urban Development. Oak Ridge, Tenn.: Oak Ridge National Laboratory.

[OTA] Office of Technology Assessment. 1982. *Energy Efficiency of Buildings in Cities*. OTA-E-168. Washington, D.C.

———. 1992. *Building Energy Efficiency*. OTA-E-518. Washington, D.C.

Padian, A. 1994a. "Confessions of an Addicted Auditor." *Home Energy* 11(3): 29–30.

———. (New York City Weatherization Coalition, Inc.). 1994b. Personal communication to Sandra Nolden.

———. 1994c. Personal communication to Tom Wilson.

Palermini, D. 1989. *Heating Cost Allocation and Conversion to Tenant Metering in Multi-Family Housing.* Portland, Ore.: City of Portland Energy Office.

Palmiter, L., and P.W. Francisco. 1994. "Measured Efficiency of Forced-Air Distribution Systems in 24 Homes." In *Proceedings of the ACEEE 1994 Summer Study on Energy Efficiency in Buildings,* 3:177–187. Washington, D.C.: American Council for an Energy-Efficient Economy.

Palmiter, L., J.R. Olson, and P.W. Francisco. 1994. *Measured Efficiency Improvements from Duct Retrofits on Six Electrically-Heated Homes.* Seattle: Ecotope, Inc.

Penn, C. 1994. "Computerized Energy Audits." *Home Energy* 11(3): 27.

Perich-Anderson, J., and L. Dethman. 1994. "How Well Is Our Energy Code Working? An Evaluation of the Tacoma, Washington, Model Conservation Program." In *Proceedings of the ACEEE 1994 Summer Study on Energy Efficiency in Buildings,* 6:149–156. Washington, D.C.: American Council for an Energy-Efficient Economy.

Perkins & Will and the Ehrenkrantz Group. 1982. *Energy Conservation for Housing: A Workbook.* Washington, D.C.: U.S. Department of Housing and Urban Development.

Perlman, M., and N. Milligan. 1988. "Hot Water and Energy Use in Apartment Buildings." *ASHRAE Transactions* 94(1).

Peterson, G. 1983. *High Efficiency Tune-up for Conversion Boilers.* MEO/TR83-1. Minneapolis: Minneapolis Energy Office.

———. 1984. "Correcting Uneven Heating in Single Pipe Steam Buildings: The Minneapolis Steam Control System." In *Proceedings of the ACEEE 1984 Summer Study on Energy Efficiency in Buildings,* 3:153–176. Washington, D.C.: American Council for an Energy-Efficient Economy.

———. 1985. *Achieving Even Space Heating in Single Pipe Steam Buildings.* MEO/TR85-8-MF. Minneapolis: Minneapolis Energy Office.

———. 1986. "The Multifamily Pilot Project: Single Pipe Steam Balancing, Hot Water Outdoor Reset." In *Proceedings of the ACEEE 1986 Summer Study on Energy Efficiency in Buildings,* 1:183–196. Washington, D.C.: American Council for an Energy-Efficient Economy.

Pigg, S. (Wisconsin Energy Conservation Corp., Madison). 1994. Personal communication to Sandra Nolden.

Poole, C. 1989. "Targeting Services by Fuel Consumption: Montana's Comprehensive Database." *Energy Exchange.* July.

Proctor, J. (Proctor Engineering Group, Corte Madera, Calif.). 1993. Personal communication to John DeCicco.

Raab, J., and A. Levine. 1981. "State and Local Programs to Encourage Multifamily Energy Conservation." In *Multifamily Energy Conservation: A Reader,* edited by S. Kaye, 61–81. Hartford, Conn.: Coalition of Northeast Municipalities.

[RECS 1984]. Energy Information Administration. 1987. *Residential Energy Consumption Survey Housing Characteristics 1984.* DOE/EIA-0314(87). Washington, D.C.: U.S. Department of Energy.

[RECS 1990a] Energy Information Administration. 1992. *Housing Characteristics 1990: Residential Energy Consumption Survey.* DOE/EIA-0314(90). Washington, D.C.: U.S. Department of Energy. May.

[RECS 1990b] Energy Information Administration. 1993a. *Household Energy Consumption and Expenditures 1990: Residential Energy Consumption Survey.* DOE/EIA-0321(90). Washington, D.C.: U.S. Department of Energy. February.

[RECS 1990c] Energy Information Administration. 1993b. *Household Energy Consumption and Expenditures 1990 Supplement: Regional, Residential Energy Consumption Survey.* DOE/EIA-0321(90)/S. Washington, D.C.: U.S. Department of Energy.

Ritschard, R., and D. Dickey. 1984. "Energy Conservation in Public Housing: It Can Work." *Energy Systems and Policy* 8(3): 269–291.

Ritschard, R., and J.A. McAllister. 1992. "Persistence of Savings in Multifamily Public Housing." In *Proceedings of the ACEEE 1992 Summer Study on Energy Efficiency in Buildings,* 4:201–215. Washington, D.C.: American Council for an Energy-Efficient Economy.

Robinson, D., G. Nelson, and R. Nevitt. 1986. *Evaluation of the Energy and Economic Performance of Twelve Multifamily Buildings Retrofitted Under a Shared Savings Program.* St. Paul, Minn.: Energy Resource Center.

———. 1988. "Evaluation of Front-End Boiler Retrofits in Two Multifamily Buildings." In *Proceedings of the ACEEE 1988 Summer Study on Energy Efficiency in Buildings,* 2:159–174. Washington, D.C.: American Council for an Energy-Efficient Economy.

Rodberg, L. 1991. *Computerization of the New York State Weatherization Assistance Program.* NYSERDA Report 91-13. Albany: New York State Energy Research and Development Authority.

————. (Queens College, Flushing, N.Y.). 1994. Personal communication to John DeCicco.

Rosenberg, M. 1984. "Metering Conversions: An Equitable and Cost-Effective Conservation Investment for Tenants?: A Massachusetts Case Study." In *What Works: Documenting Energy Conservation in Buildings,* edited by J. Harris and C. Blumstein, 161–171. Washington, D.C.: American Council for an Energy-Efficient Economy.

Sachi, M., D.L. Bohac, M.J. Hewett, T.S. Dunsworth, M. Hancock, and R.W. Landry. 1989. *Comparative Energy Performance of Domestic Hot Water Systems in Mutifamily Buildings.* MEO/TR 89-5-MF. Minneapolis: Minneapolis Energy Office.

Sachs, B. (Vermont Energy Investment Corporation, Burlington). 1993. Personal communication to Janice DeBarros.

Sachs, B., and A. Hunt. 1992. "The Critical Role of State Housing Finance Agencies in Promoting Energy Efficiency in Buildings." In *Proceedings of the ACEEE 1992 Summer Study on Energy Efficiency in Buildings,* 6:221–227. Washington, D.C.: American Council for an Energy-Efficient Economy.

Saxonis, W.P. 1993. "The Multifamily Building Evaluation Project." In *Energy Program Evaluation: Uses, Methods, and Results,* 132–138. Proceedings of the 1993 National Energy Program Evaluation Conference. Chicago.

Schlegel, J., J. McBride, S. Thomas, and P. Berkowitz. 1990. "The State-of-the-Art of Low-Income Weatherization: Past, Present, and Future." In *Proceedings of the ACEEE 1990 Summer Study on Energy Efficiency in Buildings,* 7:205–228. Washington, D.C.: American Council for an Energy-Efficient Economy.

Schuldt, M., P. Brandis, and D. Lerman. 1994. "Persistence of Savings in New Multifamily Buildings." In *Proceedings of the ACEEE 1994 Summer Study on Energy Efficiency in Buildings,* 6:185–190. Washington, D.C.: American Council for an Energy-Efficient Economy.

[SCL] Seattle City Light. 1991a. *The Multifamily Conservation Program Comprehensive Evaluation: Cost-Effectiveness Analysis.* Seattle: Seattle City Light, Energy Management Services Division.

————. 1991b. *The Multifamily Conservation Program Comprehensive Evaluation: Electricity Savings and Cost Analysis.* Seattle: Seattle City Light, Energy Management Services Division.

Sherman, M., H.E. Feustel, and D. Dickerhoff. 1989. *Description of a System for Measuring Interzonal Air Flows Using Multiple Tracer Gases.* LBL-26538. Berkeley: Lawrence Berkeley Laboratory.

Snell, J. (Citizens Conservation Corporation, Boston). 1994. Personal communication to John DeCicco.

Sonderegger, R., and J. Dixon. 1983. *CIRA—A Microcomputer Based Energy Analysis and Auditing Tool for Residential Applications.* LBL-15270. Berkeley: Lawrence Berkeley Laboratory, Energy Efficient Buildings Program.

Stiles, M.R. Forthcoming. *Development of Convective Testing Methods for Low-Rise Multifamily Dwellings.* NYSERDA report prepared by Synertech Systems Corporation. Albany: New York State Energy Research and Development Authority.

Strahs, S. 1981. "The Challenge of Financing Energy Improvements in Multifamily Housing." In *Multifamily Energy Conservation: A Reader,* edited by S. Kaye, 47–58. Hartford, Conn.: Coalition of Northeast Municipalities.

Stum, K. 1992. "Energy Savings Potential in Lighting of New Residential Dwellings." In *Proceedings of the ACEEE 1992 Summer Study on Energy Efficiency in Buildings,* 6:80–81. Washington, D.C.: American Council for an Energy-Efficient Economy.

Synertech. 1987. *Integrating Analytical Tactics into New York State's Weatherization Assistance Program: Project Findings.* Report prepared by Synertech Systems Corporation. Albany: New York State Energy Research and Development Authority.

———. 1990. *An Evaluation of the New York State Weatherization Assistance Program: Final Project Report.* Report SYN TR 90-513. Prepared by Synertech Systems Corporation. Albany: New York State Energy Office.

———. Forthcoming. *Advanced Pressure Diagnostics Methods for Characterizing Air Flows in Multifamily Buildings.* Report for NYSERDA. Prepared by Synertech Systems Corporation. Albany: New York State Energy Research and Development Authority.

SyrESCO. 1994. *Affecting Human Energy Use Through a Holistic Approach to Energy Conservation in Multifamily Buildings.* Report prepared by SyrESCO, Inc. Albany: New York State Energy Research and Development Authority.

Szydlowski, R.F., and R.C. Diamond. 1989. *Data Specification Protocol for Multifamily Buildings (ASTM Standard Practice E 1410-91).* Report LBL-27206. Berkeley: Lawrence Berkeley Laboratory. May.

Thomas, G. (SyrESCO, Inc., Syracuse, N.Y.). 1990. Personal communication to Tom Wilson.

———. 1992. "Performance Contracting: An ESCo Perspective." *Home Energy* 9(6): 23–25.

Tsongas, G. 1995. "Carbon Monoxide from Ovens: A Serious IAQ Problem." *Home Energy* 12(5): 18–21.

[UCETF/PEO] Urban Consortium Energy Task Force and Portland Energy Office. 1992. "Energy Savings from Operation and Maintenance Training for Apartment Boiler Heating Systems:

An Energy Study on Ten Low-Income Buildings." Washington, D.C.: Public Technologies, Inc.

Vine, E.L., B.K. Barnes, E. Mills, and R.L. Ritschard. 1989. "The Response of Low-Income Elderly to Tenant Incentive Programs." *Energy, The International Journal* 14(11): 677–684.

Vine, E.L., R.C. Diamond, and R.F. Szydlowski. 1987. "Domestic Hot Water Consumption in Four Low-Income Apartment Buildings." *Energy, The International Journal* 12(6): 459–467.

Wallace, J. 1985. "Case Studies of Air Leakage Effects in the Operations of High Rise Buildings." Paper presented at conference on Thermal Performance of the Exterior Envelopes of Buildings III, Clearwater, Fla. December.

Weedall, M., R. Weisenmiller, and M. Shepard, eds. 1986. *Financing Energy Conservation*. Washington, D.C.: American Council for an Energy-Efficient Economy.

Weiss, P., and M. Brown. 1989. "Computer Building Energy Analysis Software." *Home Energy* 6(5): 13–18.

Wilson, A., and J. Morrill. 1995. *Consumer Guide to Home Energy Savings*. Washington, D.C.: American Council for an Energy-Efficient Economy.

Wilson, T. 1991. "Thermostats—Good News on the Setback Front." *Home Energy* 8(1): 12–18.

Wilson, T., R. Belshe, L. Kinney, and G. Lewis. 1990. *Michigan Multifamily Weatherization Research Project Report*. Draft report prepared for Michigan Department of Labor. Fairchild, Wisc.: Residential Energy Conservation Consulting Group. December.

Woolams, J. (Conserve, Inc., New York). 1993. Personal communication to Janice DeBarros.

———. 1994. Testimony on the U.S. Federal Community Reinvestment Act Reform Proposal submitted to the Office of the Comptroller of the Currency. March 23.

Yaverbaum, L., ed. 1979. *Energy Saving by Increasing Boiler Efficiency*. Park Ridge, N.J.: Noyes Data Corporation.

About the Authors

JOHN DECICCO, a Senior Associate with ACEEE, began examining energy use in buildings in the late 1970s. He turned his attention to apartment buildings in 1983 as part of his graduate research at Princeton University's Center for Energy and Environmental Studies where his doctoral dissertation was a case study of a mid-rise public housing complex. He joined the ACEEE staff in 1990 where he also works on strategies for improving energy efficiency in transportation.

RICK DIAMOND, a Staff Scientist at Lawrence Berkeley National Laboratory, has managed research on energy use in apartment buildings for the past 15 years. He taught courses on energy and design at Harvard University's Graduate School of Design and at the California College of Arts and Crafts. He holds an A.B. from Harvard College in visual and environmental studies and an M. Arch. and Ph.D. in architecture from the University of California at Berkeley.

SANDRA L. NOLDEN is a policy analyst with Citizens Conservation Corporation (CCC) in Boston where she has conducted research under the DOE/HUD initiative on the persistence of savings from heating system retrofits in low-income apartment buildings. She co-authored a HUD-funded guidebook to aid public housing authorities in calculating utility allowances, and helped design utility DSM programs for multifamily housing. Before joining CCC in 1993, Ms. Nolden wrote for the independent weekly *California Energy Markets*; prior to that, she worked for Southern California Edison as a policy analyst and examined the issue of environmental externalities with regard to the utility's decision-making process. She holds a B.A. in economics from the University of California at Santa Barbara and an M.A. from Yale University where she focused on international environmental and natural resource concerns.

JANICE DEBARROS is Vice-President of EUA/Citizens Conservation Services, a subsidiary of EUA Cogenex Corporation formed in March, 1995 through the acquisition of staff and contracts of Citizens Conservation Corporation (CCC). While at CCC, Ms. DeBarros managed consulting efforts with HUD, Fannie Mae, three foundations, and a research project on the persistence of savings funded through the DOE/HUD initiative. Formerly Deputy Assistant Secretary for Energy Programs in the Massachusetts Executive Office of Communities and Development, Ms. DeBarros conceived, designed, and administered energy efficiency programs for publicly assisted housing utilizing oil overcharge, state funding, and Weatherization Assistance Program funds.

TOM WILSON contributed to this book as a building scientist with the Synertech Systems Corporation in Syracuse, New York. He is presently associated with Residential Energy Conservation in Fairchild, Wisconsin, a broad-based consulting firm that works with housing and energy issues with an emphasis on retrofit technologies. In 1990 he co-authored *The Michigan Multifamily Weatherization Research Project Report* with Rana Belshe; it is an extensive literature search and analysis of the state of the art in multifamily retrofit work. He was also a principal researcher and author of *Five Case Studies of Multifamily Weatherization Programs* which was prepared for Oak Ridge National Laboratory in 1995.

Index

A
A Simplified Energy Analysis Method (ASEAM) audit software, 243
acronym spellouts, 217-20
addresses
 audit software providers and manufacturers, 242, 243, 244
 resource organizations, 221-29
Affordable Housing Trust Funds, 192
air conditioning
 percentage of apartment households equipped with, 28
 rebate programs, 149-50
air flow patterns
 and air leakage sites, 101-3
 table, 79
 and door retrofits, 96-97
air leakage. *See* ventilation and air leakage
air-to-air heat pumps, 88
American Society of Heating, Refrigeration, and Air-Conditioning Engineers. See ASHRAE
apartment buildings, 1, 5-6, 15-30, 44-46
 age of buildings, 20-22
 tables, 20, 21
 air conditioning use, 28
 air flow patterns and door retrofits, 96-97
 appliances found in, 28-29
 construction standards programs, 142
 construction starts, 44
 convective bypass sites, 101-3
 table, 79
 definition, 15
 demographics, 45-46
 distribution in U.S., 18-20
 age of buildings, 20-22

maps, 19
 tables, 18, 20
 urban, suburban, and rural distributions (table), 20
 evaluating conservation program savings, 157-58
 floor area, 22
 heating fuels used, 25-28
 insulation levels, 30
 loss to future expiring federal mortgage subsidies, 175-76
 number in U.S., 16-18
 physical condition, 24-25
 rent increase trends, 44-45
 rental versus ownership, 20, 21
 size, 18, 22
 thermal mass effect on thermostat systems, 66
 vacancy rates, 45
 water heating fuel types, 28
 See also ownership of apartment buildings; recommended policies and tactics; retrofits; shell retrofits; tenants; ventilation and air leakage; weatherization programs
apartment retrofit programs, 123-74
 in efficiency programs, 123-24, 202, 215-16
 research needs, 212-14
 utility program underrepresentation, 29, 211
 weatherization program underrepresentation, 126-29
 See also government conservation programs; private-sector conservation programs; utility conservation programs; weatherization programs

tax credits, 176, 178-80
  Energy Tax Act, 178
  Low Income Housing Preservation
    and Resident Homeownership
    Act (LIHPRHA), 179-80, 191
tax disincentives, 10
tax incentives, 180, 209
underwriting policy adaptations, 200
fire doors as retrofits, 96
5220 South Drexel window retrofit. *See*
  Chicago complex window retrofit
fluorescent lighting retrofits, 119
  See also compact fluorescent (CFL)
    retrofits
free-driver (definition), 161
free-rider (definition), 161
front-end boiler retrofits, 74-76, 113-14
  for water heating systems, 111, 113-14
fuel consumption
  tracking, 57
  types of fuel used, 25-28
fuel oil
  consumption and expenditure for U.S.
    apartment households (table), 35
  oil overcharge funds as a means of
    financing energy efficiency
    improvements, 189-90
fuels. *See* electricity; fuel oil; natural gas
furnaces, electric, 87

G
gas. *See* natural gas
gas heating. *See* natural gas heating
geographical factors in energy
  consumption. *See* regional variation
Gill-Polley, Lydia, on setback thermostat
  education, 66
Goldman conservation potential study,
  40, 48, 55-56
government conservation programs
  evaluating, 158-59
  financing alternatives, 131-32
  insulation programs, 98-99
  need for apartment programs, 124
  recommended policies, 203-9, 215-16
  refrigerator efficiency programs, 117-18
  See also DOE; federal conservation
    programs; financing energy
    efficiency improvements; HUD;
    local and regional conservation
    programs; private-sector
    conservation programs; state
    conservation programs;
    weatherization programs
grants as a means of financing energy

efficiency improvements, 176, 177
gross savings (definition), 161
groundwater heat pumps (electric), 88-89

H
hazards
  of air leak sealing, 103
  of heating with kitchen stoves, 118
  of lowering water temperatures, 109
  of window retrofits, 92
heat pumps (electric), 87
  air-to-air heat pumps, 88
  groundwater heat pumps, 88-89
  HUD groundsource heat pump project,
    88-89
  as a replacement for ducted systems,
    88-89
  thermostat setbacks and, 88
heating, 63-66
  cost disclosure ordinances, 142
  electric heating distribution in U.S.
    apartment buildings, 25-27
  fuel types used, 25-28
  kitchen stoves as supplemental
    heating, 118
  natural gas heating distribution in U.S.
    apartment buildings, 25-26
  rebate programs, 149-50
  See also boiler-based heating systems;
    electric heating systems; heating
    system energy conservation
    techniques; single-pipe steam
    heating systems; thermostatic
    controls; two-pipe steam heating
    systems; water heating
heating cost allocation, 11, 61
  as a barrier to retrofits, 11
  monitoring systems, 61, 62
  recommended research on, 214
  savings, 62
  standards for, 61
heating cost disclosure ordinances, 142
heating degree day (definition), 161
heating system energy conservation
    techniques
  air leakage effect on, 101
  equity issues, 61, 62-63
  individual tenant billing effect on, 60,
    84
  shell retrofits as, 90
  unit-level metering, 62-63
  vent dampers, 71-72
  window retrofits combined with, 89,
    91, 95-96